Springer Theses

Recognizing Outstanding Ph.D. Research

For further volumes:
http://www.springer.com/series/8790

Aims and Scope

The series "Springer Theses" brings together a selection of the very best Ph.D. theses from around the world and across the physical sciences. Nominated and endorsed by two recognized specialists, each published volume has been selected for its scientific excellence and the high impact of its contents for the pertinent field of research. For greater accessibility to non-specialists, the published versions include an extended introduction, as well as a foreword by the student's supervisor explaining the special relevance of the work for the field. As a whole, the series will provide a valuable resource both for newcomers to the research fields described, and for other scientists seeking detailed background information on special questions. Finally, it provides an accredited documentation of the valuable contributions made by today's younger generation of scientists.

Theses are accepted into the series by invited nomination only and must fulfill all of the following criteria

- They must be written in good English.
- The topic should fall within the confines of Chemistry, Physics, Earth Sciences and related interdisciplinary fields such as Materials, Nanoscience, Chemical Engineering, Complex Systems and Biophysics.
- The work reported in the thesis must represent a significant scientific advance.
- If the thesis includes previously published material, permission to reproduce this must be gained from the respective copyright holder.
- They must have been examined and passed during the 12 months prior to nomination.
- Each thesis should include a foreword by the supervisor outlining the significance of its content.
- The theses should have a clearly defined structure including an introduction accessible to scientists not expert in that particular field.

Carola S. Vogel

High- and Low-Valent *tris*-N-Heterocyclic Carbene Iron Complexes

A Study of Molecular and Electronic Structure

Doctoral Thesis accepted by
The Friedrich-Alexander-University,
Erlangen-Nuremberg, Germany

Author
Dr. Carola S. Vogel
Department of Chemistry and Pharmacy
Inorganic Chemistry
Friedrich-Alexander-University
  Erlangen-Nuremberg, Egerlandstraße 1
91058 Erlangen
Germany

Supervisor
Prof. Dr. Karsten Meyer
Department of Chemistry and Pharmacy
Inorganic Chemistry
Friedrich-Alexander-University
  Erlangen-Nuremberg, Egerlandstraße 1
91058 Erlangen
Germany

ISSN 2190-5053
ISBN 978-3-642-27253-0
DOI 10.1007/978-3-642-27254-7
Springer Heidelberg New York Dordrecht London

e-ISSN 2190-5061
e-ISBN 978-3-642-27254-7

Library of Congress Control Number: 2011945160

© Springer-Verlag Berlin Heidelberg 2012
This work is subject to copyright. All rights are reserved by the Publisher, whether the whole or part of the material is concerned, specifically the rights of translation, reprinting, reuse of illustrations, recitation, broadcasting, reproduction on microfilms or in any other physical way, and transmission or information storage and retrieval, electronic adaptation, computer software, or by similar or dissimilar methodology now known or hereafter developed. Exempted from this legal reservation are brief excerpts in connection with reviews or scholarly analysis or material supplied specifically for the purpose of being entered and executed on a computer system, for exclusive use by the purchaser of the work. Duplication of this publication or parts thereof is permitted only under the provisions of the Copyright Law of the Publisher's location, in its current version, and permission for use must always be obtained from Springer. Permissions for use may be obtained through Rights Link at the Copyright Clearance Center. Violations are liable to prosecution under the respective Copyright Law.
The use of general descriptive names, registered names, trademarks, service marks, etc. in this publication does not imply, even in the absence of a specific statement, that such names are exempt from the relevant protective laws and regulations and therefore free for general use.
While the advice and information in this book are believed to be true and accurate at the date of publication, neither the authors nor the editors nor the publisher can accept any legal responsibility for any errors or omissions that may be made. The publisher makes no warranty, express or implied, with respect to the material contained herein.

Printed on acid-free paper

Springer is part of Springer Science+Business Media (www.springer.com)

Parts of this thesis have been published in the following journal articles:

J.J. Scepaniak, C.S. Vogel, M.M. Khusniyarov, F.W. Heinemann, K. Meyer, J.M. Smith, Synthesis, structure, and reactivity of an iron(V) nitride. Science **331**, 1049–1052 (2011)

J.J. Scepaniak, T.D. Harris, C.S. Vogel, J. Sutter, K. Meyer, J.M. Smith, Spin crossover in a four-coordinate iron(II) complex. J. Am. Chem. Soc. **133**, 3824–3827 (2011)

C.S. Vogel, F.W. Heinemann, M.M. Khusniyarov, K. Meyer, Unexpected reactivity resulting from modifications of the ligand periphery: Synthesis, structure, and spectroscopic properties of iron complexes of new tripodal N-heterocyclic carbene (NHC) ligands, Inorganica Chimica Acta **364**, 226–237 (2010) (invited article to special issue in honor of Prof. Dr. A.L. Rheingold)

C.S. Vogel, F.W. Heinemann, J. Sutter, C. Anton, K. Meyer, An iron nitride complex. Angew. Chem. Int. Ed. **47**, 2681–2684 (2008)

Supervisor's Foreword

Nitrogen fixation is the natural process, by which atmospheric dinitrogen is converted to ammonia at the active site of the iron/molybdenum-containing enzyme nitrogenase. This process is pivotal because fixed nitrogen is required for the biosynthesis of the basic building blocks of life. Industrially, ammonia is produced in the Haber–Bosch process by direct hydrogenation of dinitrogen with dihydrogen gas; a reaction which is catalyzed by a heterogeneous iron catalyst. However, in contrast to the ambient conditions of the biological parent system, the Haber–Bosch process requires temperatures of around 450 °C and pressures of 300 bar, thereby consuming vast amounts of energy (1.5% total world energy production). The conversion process, that has been shown to involve a surface iron nitride reactive intermediate, is still subject to intense fundamental studies, as structural and mechanistic insights that could lead to a more energy efficient synthesis of ammonia are highly desirable. As a result, well-characterized high-valent iron complexes have been sought as biomimetic models for transformations mediated by iron-containing enzymes. To gain understanding of iron nitride reactivity, and the possible role of such species in catalysis, insight into the molecular and electronic structure of complexes stabilizing the iron nitride synthon is essential. In this context, iron nitride species and their possible role in catalysis have recently gained increasing attention. However, high-valent iron nitride species have proven very challenging to isolate and characterize. Chelating N-heterocyclic carbene ligands that enforce tripodal topologies at coordinated metal centers have the possibility to stabilize metal centers in low and high oxidation states, including the iron nitrido moiety; and thus, provide powerful platforms for the interrogation of the molecular and electronic structure.

In the course of her doctoral research, Carola Vogel, in collaboration with an international team of scientists, succeeded in the synthesis and the structural and spectroscopic characterization of the first isolable high-valent Fe(IV) and Fe(V) nitrido complexes. These complexes, stabilized by tripodal N-heterocyclic carbene ligands, were obtained *via* photolysis of the corresponding iron azido complexes. Reactivity studies demonstrated the reaction of a unique iron(V) nitrido complex with water to yield almost quantitative yields of ammonia with subsequent formation of a well-defined iron(II) species *under ambient conditions*; thus, providing the possibility for closing a less energy intensive synthetic cycle for the production of ammonia. All new iron complexes have been characterized in detail by standard methods, as well as single crystal X-ray structure determination, SQUID magnetization measurements, and EPR- and Mößbauer spectroscopy. Despite these exciting new discoveries, considerable challenges remain, including the direct synthesis of an iron nitride from dinitrogen.

Prof. Dr. Karsten Meyer

Acknowledgments

I particularly would like to thank my supervisor, Prof. Dr. Karsten Meyer, for giving me the opportunity to undertake my Ph.D. studies at Friedrich-Alexander-University Erlangen-Nuremberg. I would also like to thank Prof. Meyer providing me with a highly interesting research topic. In addition I would like to acknowledge the possibility to conduct my research in a way that allowed me to develop my own ideas and add my own creativity to this project.

I would also like to thank Jeremiah Scepaniak and Prof. Dr. Jeremy Smith (New Mexico State University). I am very grateful for the fruitful cooperation which so far has resulted in two exciting publications.

Frau Christina Wronna is acknowledged for performing the elemental analyses.

The help of Dr. Matthias Moll in recording a large number of NMR spectra is highly appreciated.

Dr. Eckhard Bill (MPI Mülheim) is acknowledged for his help in recording and interpreting several Mößbauer spectra.

I would like to thank Dr. Jörg Sutter for his help in recording and interpreting a number of EPR and Mößbauer spectra.

Dr. Marat Khusniyarov is acknowledged for conducting a number of theoretical calculations and in addition for his support in recording and interpreting several EPR and SQUID measurements.

I would like to thank Dr. Carsten Streb for his help and support during the writing phase of my thesis.

In addition, I would like to thank all the members of the Meyer group for their help in the lab and for the friendly atmosphere.

The help of Dr. Frank Heinemann in solving and refining a large number of crystallographic datasets is highly appreciated, as they form a substantial part of this thesis.

Finally and most importantly, I would like to thank my family and friends, particularly my parents and my husband Frank for their endless love and support.

Dr. Carola S. Vogel

Contents

**1 Introduction to Tripodal N-Heterocyclic Carbene
Iron Complexes** 1
  1.1 Carbenes 1
  1.2 History of N-Heterocyclic Carbenes (NHC)
      and Their Application 3
  1.3 NHC Ligands for the Stabilization of
      Low and High Oxidation State Transition Metals 5
  1.4 Types of Tripodal Ligand Environments 6
      1.4.1 Overview of Tripodal Ligands 6
      1.4.2 Tripodal N-Heterocyclic Carbene Ligands 8
  1.5 High-Valent Iron Species: Intermediates in Enzymatic
      Reactions and the Haber–Bosch Process. 9
  1.6 Objectives. 10
  1.7 Summary and Acknowledgements 12
  References 14

2 TIMEN^{mes}: An Iron Nitride Complex 19
  2.1 Introduction 19
  2.2 Results and Discussion. 20
      2.2.1 General Synthetic Procedure for the Precursors
             of the TIMEN^R System 20
      2.2.2 The TIMEN^{mes} Ligand and its Coordination to Iron:
             Synthesis Towards an Elusive Iron Nitride Complex 21
  2.3 Conclusion 40
  2.4 Experimental. 41
      2.4.1 Methods, Procedures, and Starting Materials 41
      2.4.2 Computational Details 42
      2.4.3 Synthetic Details. 43
      2.4.4 X-ray Crystal Structure Determination Details 45
  References 50

3 TIMEN$^{tol/3,5xyl}$: Unexpected Reactivity Resulting From Modifications of the Ligand Periphery

3.1	Introduction	53
3.2	Results and Discussion	55
3.3	Theoretical Considerations	69
3.4	Conclusion	71
3.5	Experimental	74
	3.5.1 Methods, Procedures, and Starting Materials	74
	3.5.2 Computational Details	75
	3.5.3 Synthetic Details	75
	3.5.4 X-ray Crystal Structure Determination Details	81
References		82

4 TIMEN^{3,5CF3}: Hybrid System Capable of Cyclometallation and Nitride Insertion

4.1	Introduction	85
4.2	Results and Discussion	86
4.3	Conclusion	92
4.4	Experimental	93
	4.4.1 Methods, Procedures, and Starting Materials	93
	4.4.2 Synthetic Details	94
	4.4.3 X-ray Crystal Structure Determination Details	96
References		100

5 PhB(^{tBu}Im)$_3^-$: Synthesis, Structure, and Reactivity of an Iron(V) Nitride

5.1	Introduction	101
5.2	Results and Discussion	102
5.3	Conclusion	112
5.4	Experimental	112
	5.4.1 Methods, Procedures, and Starting Materials	112
	5.4.2 Computational Details	113
	5.4.3 Synthetic Details	113
	5.4.4 X-ray Crystal Structure Determination Details	115
References		117

6 PhB(^{Mes}Im)$_3^-$: Spin Crossover in a Four-Coordinate Iron(II) Complex

6.1	Introduction	119
6.2	Results and Discussion	120
6.3	Conclusion	125

6.4	Experimental	126
	6.4.1 Methods, Procedures, and Starting Materials	126
	6.4.2 Synthetic Details	127
	6.4.3 Magnetic Susceptibility Measurements	128
	6.4.4 Computational Details	129
	6.4.5 X-ray Crystal Structure Determination Details	130
References		132

Symbols and Abbreviations

Å	Ångstrøm; 10^{-10} m
Å^3	Cubic Ångstrøm
acetonitrile-d_3	Deuterated acetonitrile
Ad	Adamantane
$\angle$	Angle
av	Average
β	Beta
benzene-d_6	Deuterated benzene
B.M.	Bohr Magneton (1 B.M. = 9.27400915(23) $\times$ 10^{-24} JT^{-1})
B3LYP	Becke 3-parameter Lee–Yang–Parr exchange-correlation functional
BP86	Becke's 88 exchange functional and Perdew's 86 correlation
br	Broad
°C	Degree Celsius
CCDC	Cambridge crystallographic data centre
CD_2Cl_2	Deuterated methylene chloride
CH_2Cl_2	Methylene chloride
CH_3CN	Acetonitrile
calcd.	Calculated
cm	Centimeter
Cp	Cyclopentadienyl
CS	Closed-shell
CT	Charge Transfer
cyclam	1,4,8,11-tetraazacyclotetradecane
CW-EPR	Continuous wave electron paramagnetic resonance
°	Degree
δ	NMR chemical shift; ppm
δ	Mößbauer isomer shift; mm s^{-1}

Δ	Difference
δ	Orbital symmetry
d	Distance
d	Doublet
D	Zero-field splitting parameter
DCM	Double crystal monochromator
ΔE_Q	Mößbauer quadrupole splitting parameter; mm s^{-1}
DFT	Density functional theory
$\Delta H_{1/2}$	Electron paramagnetic resonance line width; mT
DMSO	Dimethyl sulfoxide
DMSO-d_6	Deuterated dimethyl sulfoxide
E	Energy
e$^-$	Electron
e.g.	For example, abbreviation of Latin *'exempli gratia'*
EPR	Electron paramagnetic resonance
ESI-MS	Electron spray ionization mass spectroscopy
et al.	And others, abbreviation of Latin *'et alii'*
eV	Electron volt
ε	Molar extinction coefficient; M^{-1}cm^{-1}
η	Asymmetry parameter
ζ	Spin orbit coupling constant; cm^{-1}
Et$_2$O	Diethyl ether
Fc$^+$/Fc	Ferrocenium/ferrocene
g	EPR g-value
g_e	EPR g-value of the free electron; $g_e = 2.0023$
$g_{||}$	EPR g-value parallel
$g_\perp$	EPR g-value perpendicular
Γ_{FWHM}	Line width, FWHM = full-width at half-maximum; mm s^{-1}
γ	Gamma
GHz	Gigahertz; 10^9 s^{-1}
GooF	Goodness of fitting
h	Hour
HCl	Hydrochloric acid
HOMO	Highest occupied molecular orbital
HS	High spin
i.e.	That is, abbreviation of Latin *'id est'*
IR	Infrared
J	Coupling constant, Hz
J	Total angular momentum
χ_M	Molar magnetic susceptibility; M^{-1} cm^{-1}
K	Kelvin

KBr	Potassium bromide
kJ	Kilo-Joule (1 kcal = 4.1868 kJ)
KO^{t}Bu	Potassium *tert*-butoxide
λ	Wavelength, lambda
L	Total orbital angular momentum
L_{ax}	Axial ligand
LMCT	Ligand-to-metal charge transfer
LS	Low spin
LUMO	Lowest unoccupied molecular orbital
M	Molar; mol L^{-1}
m	Multiplet
m^3	Cubic meter
MeCN	Acetonitrile
MHz	Megahertz; 10^6 s^{-1}
mes	Mesityl
mg	Milligram
μL	Microliter
mL	Milliliter
mM	Millimolar
mm	Millimol
MO	Molecular orbital
mW	Milliwatts
μ_B	Bohr magneton; $\mu B = 9.27408 \times 10^{-24}$ JT^{-1}
μ_{eff}	Effective magnetic moment
MW	Molecular weight
NHC	N-heterocyclic carbene
NHE	Normal hydrogen electrode
nm	Nanometer; 10^{-9} m
NMR	Nuclear magnetic resonance
OAc$^-$	Acetate anion
obsd.	Observed
oop	Out-of-plane shift
Oe	Oersted; A m^{-1}
ORTEP	Oak Ridge Thermal Ellipsoid Plot
OS	Open-shell
PF$_6{}^-$	Hexafluorophosphate anion
Ph	Phenyl
π	Pi; pi bond
ppm	Parts per million
ref.	Reference
RT	Room temperature

σ	Sigma; sigma bond; standard deviation
s	Singlet
S	Total spin angular momentum
SCO	Spin crossover
S_N2	Bimolecular nucleophilic substitution
SOMO	Singly occupied molecular orbital
solv	Solvent
SQUID	Superconducting quantum interference device
STOs	Slater-type orbitals
SV(P)	Split-valence basis set
T	Tesla
t	Triplet
T_C	Transition temperature
tert	Tertiary
TEMPO-H	1-Hydroxy-2,2,6,6-tetramethyl-piperidine
THF	Tetrahydrofuran
TIMEN^R	*tris*[2-(3-aryl-imidazol-2-ylidene)ethyl]amine, R = aryl
TIP	Temperature independent paramagnetism; χ_{TIP}
TMS	Trimethylsilyl
TPB	*tris*(pyrazolyl)borate
TPM	*tris*(pyrazolyl)methane
TREN	*tris*(2-aminoethyl)amine
TZV	Valence triple-zeta basis set
UV-vis/NIR	Ultra-violet/visible/near-infrared
$\tilde{v}$	Wavenumber; cm^{-1}
VT	Variable temperature
v	Velocity; $mm\ s^{-1}$
V	Volt
V	Volume
v_{as}	Asymmetric vibration
vs.	Against, abbreviation of Latin 'versus'
XRD	X-ray diffraction
ZORA	Zeroth-order regular approximation

Chapter 1
Introduction to Tripodal N-Heterocyclic Carbene Iron Complexes

1.1 Carbenes

First introduced to organic chemistry by Doering in the 1950s [1], and a decade later to organometallic chemistry by Fischer in 1964 [2], carbenes rapidly made their way into all fields of chemistry [3–6]. These intriguing species have aroused the interest of organic, inorganic, and theoretical chemists like no other single class of molecules. This is most probably due to their exceptional molecular and electronic structures, sophisticated syntheses, and diverse properties, varying from nucleophilic to electrophilic and even to the point of ambiphilic character [7, 8].

Carbenes are defined as neutral compounds possessing a divalent carbon atom in their structure with only six valence-shell electrons. This carbon atom is bound to two adjacent groups by covalent bonds. It has two nonbonding electrons, which may have antiparallel (singlet state) or parallel spins (triplet state).

The geometry at the carbene carbon can either be linear or bent. A linear geometry results in an sp-hybridized carbene center with two degenerate non-bonding orbitals (p_x and p_y). The linear geometry of carbenes CX_2 (Figs. 1.1, 1.2) is an extreme case and occurs if the substituents X are lower in electronegativity than the carbene carbon atom. The simplest possible carbene is methylene, CH_2, consisting of one carbon atom and two hydrogen atoms. Methylene is an example for a linear carbene with a triplet ground state [9]. Other examples for substituents inducing a linear geometry are lithium (Li–C–Li) or boron (B–C–B) [10]. Bending of the molecule removes the degeneracy and the carbene carbon adopts an sp^2-hybridization. The p_y orbital remains almost unchanged and is now often called p_π [11]. In contrast, the p_x orbital is stabilized by acquiring s character and is therefore called σ. Most carbenes are bent and contain an sp^2-hybridized carbon with the frontier orbitals systematically labeled as σ and p_π. Four electronic configurations can be envisioned (Fig. 1.3) [11]. The two nonbonding electrons at the sp^2-hybridized carbene carbon atom can occupy the two empty orbitals with parallel spin orientation, leading to a triplet ground state ($\sigma^1 p_\pi^1$, $^3B_1$ state).

C. S. Vogel, *High- and Low-Valent* tris-*N-Heterocyclic Carbene Iron Complexes*,
Springer Theses, DOI: 10.1007/978-3-642-27254-7_1,
© Springer-Verlag Berlin Heidelberg 2012

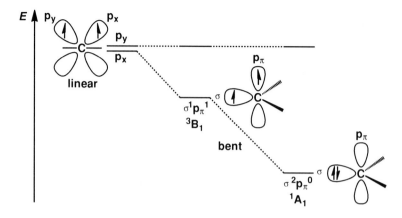

Fig. 1.1 Frontier orbitals and possible electron configurations for carbene carbon atoms [18]

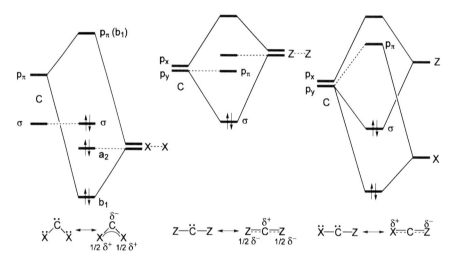

Fig. 1.2 Orbital interaction diagram showing the influence of the mesomeric effect, X: p–e- donor substituents; Z: p–e- acceptor substituents [11]

Two electrons occupying the σ orbital with antiparallel spin orientation lead to a singlet ground state ($\sigma^2 p_\pi^0$, $^1A_1$ state). There are two more electronic configurations conceivable, a less stable singlet state ($\sigma^0 p_\pi^2$, $^1A_1$ state) and an exited singlet state with an antiparallel configuration of the p_π and σ orbitals ($\sigma^1 p_\pi^1$, $^1B_1$ state) (Fig. 1.3).

In general, the multiplicity of the ground state determines the properties and reactivity of a carbene [12]. Singlet carbenes have a filled σ and an empty p_π orbital, and therefore often show an ambiphilic behaviour. Triplet carbenes can be considered as diradicals because of their two unpaired electrons. The multiplicity

1.1 Carbenes

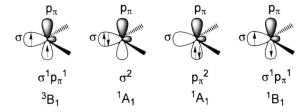

Fig. 1.3 Four possible electronic configurations for a carbene [11]

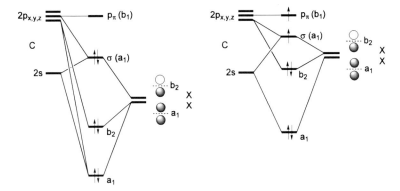

Fig. 1.4 Orbital interaction diagram showing the influence of the inductive effect [11]

of the ground state is determined by the relative energies of the σ and p_π orbitals (Fig. 1.4). A large σ–p_π separation favors the singlet state and quantum chemical calculations have shown that an energy difference of about 2 eV is required for the stabilization of the singlet ground state ($^1A_1$) [13]. Steric and electronic effects of the substituents at the carbene carbon determine the multiplicity of the ground state. The singlet ground state is stabilized by more electronegative groups, because the negative inductive effect lowers the relative energy of the nonbonding σ orbitals whereas the relative energy of the p_π orbital remains almost unchanged [14–17]. Substituents with σ-electron donating properties decrease the energy gap between the σ and p_π orbital, and thus stabilize the triplet ground state.

1.2 History of N-Heterocyclic Carbenes (NHC) and Their Application

Chemists have been searching for stable carbenes since the nineteenth century. The quest began in the year 1835 with Dumas [19]. However, his attempt to synthesize stable carbenes by treating methanol with dehydrating agents like phosphorous pentoxide or sulphuric acid failed [9], along with early experiments by Geuther

Scheme 1.1 Synthesis of the entetraamine 2=2 by α-elimination of 1

Scheme 1.2 Attempted synthesis of carbene 3 from tetraphenylimidazolium perchlorate

Scheme 1.3 Arduengo's synthesis of the first stable N-heterocyclic carbene 5

[20], Scheibler [21], and Schmeisser [22, 23]. Discussions regarding N-heterocyclic carbenes were initiated in 1960 by Wanzlick's report on the α-elimination of chloroform from **1** [24]. The postulated imidazolidin-2-ylidene **2** could never be isolated, and Wanzlick always obtained its dimer instead, the entetraamine 2=2 (Scheme 1.1).

Wanzlick's next attempt to prepare the free carbene **3** was by deprotonation of tetraphenylimidazolium perchlorate with KO′Bu (Scheme 1.2). The free carbene could not be isolated and its intermediate formation was demonstrated indirectly by identification of some of its reaction products with water or Hg(OAc)$_2$ [25].

Finally, the first stable carbene **5** was prepared, isolated, and crystallographically characterized in 1991 by Arduengo et al. [26]. The diadamantyl-substituted imidazol-2-ylidene **5** was prepared by deprotonation of the corresponding imidazolium salt **4** with sodium or potassium hydride in the presence of catalytic amounts of KO′Bu or DMSO (Scheme 1.3).

A large number of N-heterocyclic carbenes (NHCs) and acyclic heteroatom-substituted free carbenes were isolated in the last 20 years. The number of complexes with NHC ligands is even larger, some of which show remarkable catalytic properties. Numerous metal complexes bearing NHC ligands are applied as catalysts in various catalytic reactions. Catalytic applications of NHC complexes have been reviewed by Herrmann [27], Crudden [28], Nolan [29], and Glorius [30]. Particularly, NHC complexes catalyze C–C cross-coupling reactions [31, 32] and the use of carbene complexes as catalysts in olefin metathesis has been the subject of comprehensive review articles over the last years [33–35]. The excellent performance of NHC catalysts in catalytic reactions derives not exclusively from the pronounced σ-donor properties of NHC ligands but also from the possibility of π-back-bonding within the metal-carbene unit [36]. NHCs stabilize both high and low oxidation states formed in catalysis resulting in higher turnover numbers and longer catalyst lifetimes [37]. During one run of the catalytic cycle the catalyst, i.e. the NHC metal complex, undergoes reduction and oxidation steps, e.g. a two-electron oxidation of the metal center, $M^{n+} \to M^{(n+2)+}$. While the π-acceptor capability stabilizes the M^{n+}-state, the $M^{(n+2)+}$-state is supported by the σ-donor properties of the NHC ligand. Other advantages of NHC complexes, in contrast to complexes bearing phosphane and phosphite ligands, are high thermal durability, hydrolytic stability, and the reactions can often be carried out in air. A basic functional principle in homogeneous complex catalysis is based on the fact that phosphane and phosphite ligands not only protect low-valent metal centers from aggregation, but also create coordination sites in dissociation equilibria at which the catalytic elementary steps proceed. Generally, as a result of the notorious phosphane degradation by P–C bond cleavage, an excess of the ligand often 100 times more than the metal is required to control the dissociation equilibrium in homogeneous catalysis. In contrast, using N-heterocyclic carbene complexes as catalyst, the NHC ligands do not undergo deactivating degradation reactions, thus there is no need for a ligand excess [38].

1.3 NHC Ligands for the Stabilization of Low and High Oxidation State Transition Metals

In the last few years N-heterocyclic carbenes (NHC) have become an alternative to the omnipresent phosphane ligands in transition metal chemistry, establishing itself as an effective class of ligands in organometallics and catalysis [11, 30, 39–41]. Many examples from different research groups substantiate the ability of NHCs to coordinate to a diversity of metal centers and to effectively stabilize the resulting organometallic compound. Using this approach to stabilize organometallic species enables a thorough characterization and analysis of reactive intermediates that have not been observed in analogous phosphane chemistry.

Until recently, NHC ligands were considered as mainly σ-donors, while their π-acceptor properties were disregarded, even though the carbene carbon atom has a formally empty p_π orbital. Presently, it is accepted that NHC ligands have π-acceptor

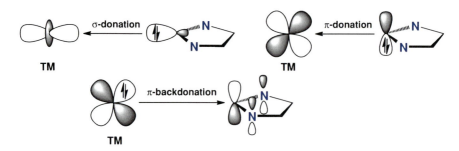

Fig. 1.5 Representation of the interaction between the NHC ligand and the transition metal (TM) fragment in terms of the Dewar–Chatt–Duncanson model

possibilities [42–48], particularly in electron-deficient metal complexes [49]. Recently, Cavallo and co-workers reviewed the research in this field to understand the nature of the transition metal (TM)-NHC bond [50]. Figure 1.5 shows the most dominant interactions between NHC ligands and transition metals. The metal–ligand bonding in transition metal complexes is typically discussed using the Dewar–Chatt–Duncanson donor–acceptor model [51]. This model separates the interaction into ligand to metal σ-donation and metal to ligand π-back-donation.

1.4 Types of Tripodal Ligand Environments

1.4.1 Overview of Tripodal Ligands

Tripodal ligand systems have been extensively used in the formation of complexes involving metals from across the periodic table [52–55]. Their enhanced chelate effect, relative to mono- and bidentate ligands, allows them to robustly ligate metal centres, and often permits the stabilization and isolation of unusual complex types and reactive intermediates. Consequently, metal complexes of tripodal ligands have found numerous applications in catalysis, enzyme mimicry, and small molecule activation [52–56]. Given their importance, the development of facile, high yield synthetic routes to new tripodal ligand systems with variable steric and electronic properties has been an ongoing theme in inorganic and organometallic coordination chemistry.

Well-known examples of tripodal ligands are the *tris*(pyrazolyl)borate ligand (TPB, **6**, Chart 1.1) and the *tris*(pyrazolyl)methane ligand (TPM, **7**, Chart 1.1). The *tris*(pyrazolyl)borate ligand has been widely used in organometallic and coordination chemistry since the initial development by Trofimenko in the late 1960s [57]. The *tris*(pyrazolyl)borate based ligands are very attractive because they coordinate strongly to early transition metals in a tridentate fashion, and the steric and electronic properties of the pyrazolyl donor can be modified by changing

1.4 Types of Tripodal Ligand Environments

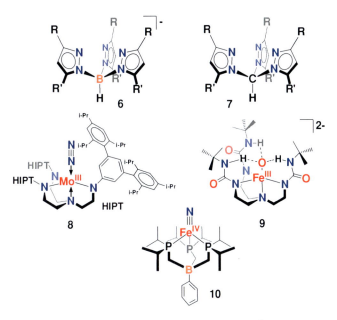

Chart 1.1 Representative tripodal ligands and complexes (R, R′ = various alkyl and aryl substituents)

its 3- and 5-substituents [58, 59]. Also, the isoelectronic *tris*(pyrazolyl)methane ligand has raised interest in the field of coordination chemistry, though its synthesis is more challenging. Due to their shape, pre-organized structure, and relatively facile synthetic access, numerous derivatives of the *tris*(pyrazolyl)borate and the *tris*(pyrazolyl)methane ligands were developed by functionalization of the pyrazolyl rings, some of them being used as artificial enzymes [60–63]. The *tris*-(2-aminoethyl)amine (TREN)-based ligands are also very popular and effective. Schrock et al. [64–66] isolated and characterized many intermediates in a hypothetical dinitrogen reduction reaction . An example of these types of molybdenum coordination compounds, containing bulky aryl-substituted *tris*(amido)amine ligands, is complex **8** (Chart 1.1). Utilizing an ureayl-functionalized TREN ligand, Borovik et al. prepared a terminal Fe(III) oxo complex **9** (Chart 1.1) [67, 68]. Peters et al. showed that a tripodal ligand environment is advantageous for stabilizing elusive species. This can be ascribed to the unique geometric and electronic properties enforced by the threefold symmetric ligand environment [69, 70]. Using an anionic *tris*(phosphane) ligand, the group has synthesized the first cobalt(III) [69] and iron(III) [71] terminal imido complexes and has synthesized and spectroscopically characterized the first iron(IV) nitride species **10** (Chart 1.1) [70, 72]. Chart 1.1 gives an overview of representative tripodal ligands and corresponding complexes.

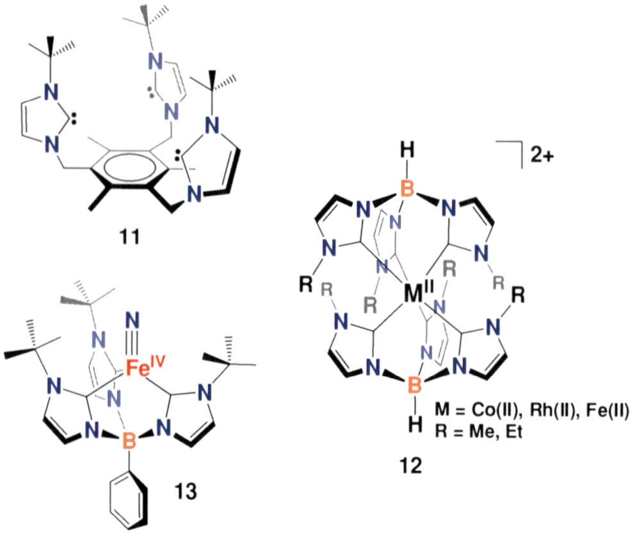

Chart 1.2 Representative tripodal N-heterocyclic carbene ligands and complexes

1.4.2 Tripodal N-Heterocyclic Carbene Ligands

Ligands containing three or more carbene centers are still rare. Shortly after Arduengo, Dias and Jin synthesized and isolated the first free *tris*(carbene) **11** but no metal complexation could be achieved [73]. Meyer et al. thoroughly investigated the coordination chemistry of **11** and its derivatives. The cavity generated by this ligand can only host exceptionally large metal ions, such as the thallium(I) cation [74]. Fehlhammer et al. introduced the carbene analogues of the *tris*(pyrazolyl)borate ligands [75]. The free carbenes could not be isolated, but Co(III), Rh(III) and Fe(III) complexes were synthesized via salt metathesis of their lithium complexes [76]. The anionic *tris*(carbene)borato ligands tend to form hexacarbene metal complexes **12** (see Chart 1.1). This is likely due to a lack of steric hindrance at the three-position of the imidazol-2-ylidene ring. The resulting complexes are coordinatively saturated and show no further reactivity [77]. This characteristic greatly limits the application of this class of ligands in metal-assisted small molecule activation. However, Smith et al. demonstrated that one can influence this feature by introducing sterically demanding substituents at the boron anchor and especially at the carbene units. By derivatizing the NHC with aryl groups like mesityl (PhB(^{Mes}Im)$_3^-$) [78], or bulky alkyl substituents, e.g. *tert*-butyl [79], Smith et al. was able to obtain *mono*-chelated transition metal complexes **13**. This modified ligand system is now similar to the ligands used in the Meyer group and the resulting complexes show similar features. Chart 1.2 gives an overview of some representative tripodal N-heterocyclic carbene ligands and complexes.

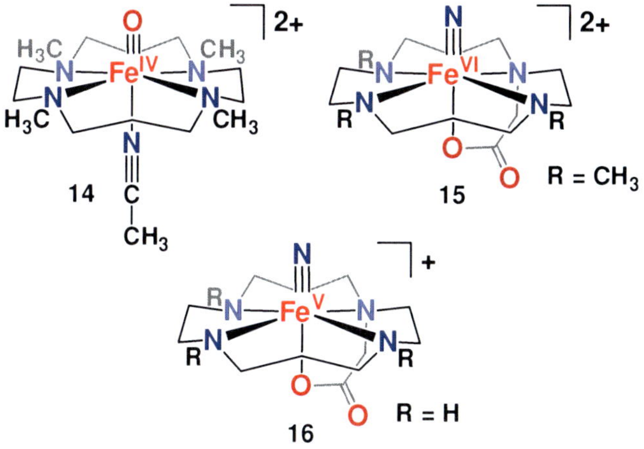

1.4 Types of Tripodal Ligand Environments

Chart 1.3 High-valent iron oxo and nitrido complexes derived from derivatized cyclam ligands

1.5 High-Valent Iron Species: Intermediates in Enzymatic Reactions and the Haber–Bosch Process

High-valent iron species are proposed as active intermediates in the cycles of many important biocatalysts. Iron(IV) is the most readily accessible high oxidation state; however, iron(V) has also been proposed as a key intermediate in some non-heme dioxygenases [80]. Very often the investigation of well designed and defined model complexes is absolutely essential to gain first insights into structural and spectroscopic properties of such intermediates. The iron(IV) oxo moiety, which has long been known to be at the catalytic centers of oxygenases [81], was first crystallographically characterized in a square pyramidal iron complex of an N-methylated 1,4,8,11-tetraazacyclotetradecane (cyclam) macrocyclic ligand (see Chart 1.3, **14**) [82]. Meyer et al. demonstrated that generation of a nitridoiron(V) species for spectroscopic characterization and to study the electronic structure is possible by employing this 1,4,8,11-tetraazacyclotetradecane (cyclam) ligand [83]. Related cyclam derivatives have also allowed the preparation and detailed spectroscopic characterization of octahedral Fe(V) [83–85] and Fe(VI) [86] nitrido complexes (see Chart 1.3, **15**, **16**). However, the structural characterization and reactivity of these fleeting intermediates, which are usually studied at low temperatures in frozen matrices, remains elusive.

Iron nitrides have also been proposed to be key intermediates in the industrial (Haber–Bosch process) [87] and biological (nitrogenase) [72, 88] synthesis of ammonia. While iron bound "surface nitride" species are observed on the catalyst surface in the Haber–Bosch process [89], recent X-ray diffraction studies on nitrogenase suggest an interstitial atom in the center of the iron-sulfur cluster of the FeMo co-factor [90]. Although it is tempting to suggest a nitride anion at the site of biological nitrogen reduction, the nature of this atom is controversial and

10   1 Introduction to Tripodal N-Heterocyclic Carbene Iron Complexes

still under debate [91, 92]. Accordingly, the synthesis and characterization of model complexes is critical to predict the reactivity of iron nitrides and their possible role in ammonia synthesis. In addition to the molecular and electronic structural insight provided by complexes that stabilize the [Fe≡N] unit, their reactivity will also impact our understanding of both biological and industrial NH_3 syntheses.

1.6 Objectives

The goal of this dissertation work was to use a tripodal N-heterocyclic carbene environment for the generation and isolation of high-valent iron species, followed by a thorough investigation of the resulting complexes to gain insight into the molecular and electronic structure and, most importantly, to study the reaction behaviour of such compounds. The main focus was on the synthesis of iron nitrido complexes and their subsequent characterization using a range of spectroscopic and analytical methods such as EPR and Mößbauer spectroscopy. In particular, the knowledge of the molecular structure and thus the exact knowledge of the atomic coordinates of the investigated molecule obtained by X-ray crystal structure determination will help to gain a deeper insight in the electronic structure via DFT (density functional theory) calculations. Good initial estimates for atomic coordinates, in the best case derived from crystal structure determination, will greatly increase the speed of DFT calculations and in many cases make a decisive difference in weather the calculations can even converge to an energy minimum [93]. Furthermore, spectroscopic data can be better simulated and hence evaluated. New insights in the molecular and electronic structure and the resulting information about reaction behaviour could help to understand the role of iron nitridos in biological reactions and potentially lead to improvements of industrial processes. Tripodal N-heterocyclic carbene ligands are in many ways highly promising candidates for these challenging tasks. They are capable of stabilizing low to high metal oxidation states. Furthermore, this ligand class is easily customizable by introducing sterically demanding substituents perpendicular to the *tris*-carbene plane forming a cavity which can host an axial ligand. Thus, even reactive ligands like a terminal nitride can be expected to be protected from decomposition and consequently be stabilized (Fig. 1.6).

One of the main reasons to choose a tripodal ligand system becomes particularly obvious when comparing the d-orbital splitting in a tetragonal and a trigonal ligand field. In tetragonal symmetry high-valent metal complexes with strongly π-donating ligands, e.g. the nitrido ligand, posses a so-called $1 + 2 + 1 + 1$ d-orbital splitting (see Fig. 1.7c) [94]. The splitting pattern was originally introduced by Ballhausen and Gray for chromyl and molybdenyl complexes, and expecially for the vanadyl complex $[OV(OH_2)_5]^{2+}$ [95]. The electrostatic model for the hydrated vanadyl ion consists of V^{4+} situated in a tetragonal electric field caused by the oxide ion and five water dipoles. The scheme of terms has also been

1.6 Objectives

Fig. 1.6 Tripodal NHC ligand system

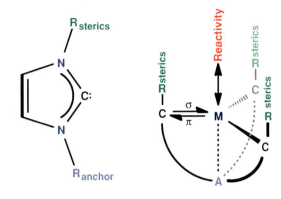

used for *trans*-dioxo and nitride complexes of Tc(V), Re(V), and Os(VI) [96, 97]. The $1 + 1 + 2 + 1$ d-orbital splitting scheme of $[Cr(N)Cl_4]^{2-}$ (see Fig. 1.7d) is distinctly different [98] from that applicable to vanadyl complexes and to other chromium and manganese nitrido complexes with stronger field equatorial ligands [99–102]. But both have in common that in fourfold symmetry only one non-bonding orbital is available whereas for a d^3 configuration, highly destabilized anti-bonding orbitals must be occupied. Accordingly, for Fe(V) nitrido complexes with a d^3 electron configuration, generation and stabilization is difficult in a tetragonal ligand environment, hence molecular iron nitride complexes could only be generated in frozen matrices [83, 86, 103]. Peters et al. could demonstrate that tripodal (phosphino)borate ligands can reveal novel aspects in coordination chemistry. In particular, the *tris*(phosphino)borate ligands $[PhBP_3^{ph}]^-$ and $[PhBP_3^{ipr}]^-$ have highlighted fundamentally new aspects in the coordination chemistry of both cobalt and iron. These ligands have stabilized the first examples of mononuclear terminal imides ($M \equiv N_x$) and nitrides ($M \equiv N$) of cobalt and iron [69, 70, 104–107]. The ligands are structurally similar to other tripodal ligands, e.g. the *tris*(carbenes) from the Meyer group and *tris*(carbene)borates from the Smith group. The qualitative frontier orbital diagrams sketched in Fig. 1.7a for a metal imide ($M \equiv N_x$) [69, 108] and in Fig. 1.7b for a metal nitrido ($M \equiv N$) [70] species, supported by a tripodal (phosphino)borate ligand, illustrate that in threefold symmetry, at least two non-bonding orbitals are available.

Compared to the imide, the nitride complex has a different orbital splitting diagram in which the d_{z^2} orbital is strongly destabilized by increased antibonding interactions with the nitrido ligand. The stability of the electronic configurations for imide and nitride species suggests that also with a tripodal *tris*(carbene) ligand an iron nitride complex should be electronically accessible.

Another topic of this thesis was the development of new tripodal N-heterocyclic ligands and their corresponding complexes with the intention of applying them in small molecule activation and atom/group transfer chemistry. In contrast to the aforementioned goal—stabilization of a reactive axial ligand—in this context another strategy has to be pursued. In order to stabilize reactive intermediates,

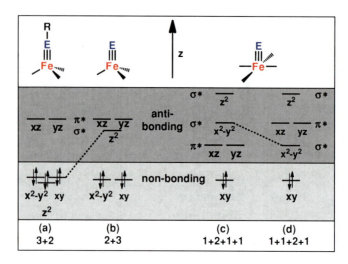

Fig. 1.7 *d*-Orbital splitting in complexes with threefold (**a** and **b**) versus fourfold symmetry (**c** and **d**)

a sterically demanding ligand is required to shield the reactive unit and to prevent the group from undergoing bimolecular decomposition via dimerization, for instance. But in terms of small molecule activation and atom/group transfer, the generation of complexes that undergo further reactions is the desired objective. Therefore, the steric hindrance of the ligand has to be reduced to give access and enable the activation and transfer of atoms or groups. Consequently, the aim must be derivatization of the established ligands by introducing sterically less demanding substituents. By changing the ligand periphery, it was hoped to synthesize new complexes which show diverse features, especially in terms of varied reactivity of the resulting compounds.

1.7 Summary and Acknowledgements

The text, schemes, and figures of Chaps. 2, 3, 5, and 6, in part or full, are reprints of the materials previously published, as listed in the publication list at page XIII.

The dissertation author was the primary researcher and author in Chaps. 2 and 3.

Chapter 2 deals with iron complexes, supported by the tripodal N-heterocyclic carbene ligand, TIMEN[mes], which was developed in the research group of Prof. Dr. Karsten Meyer. A series of iron complexes, in different oxidation states, could be synthesized bearing the TIMEN[mes] ligand. In particular, this ligand system enables the synthesis of an iron(IV) nitride, which could be fully characterized by elemental analysis, NMR-, IR-, and UV-vis spectroscopy, as well as X-ray crystal structure determination. Attempts to oxidize the iron(IV) nitride, aiming for an

1.7 Summary and Acknowledgements

iron(V) nitride, yields an insertion product, where the nitride ligand has inserted into one of the iron–carbene bonds [109].

Chapter 3 presents the synthesis of two new derivatives of the TIMEN^R family, TIMEN^{tol} and TIMEN$^{3, 5xyl}$ and their corresponding iron complexes. These new ligands were designed with the goal to reduce the steric demand of the tripodal carbene, in order to allow side access for substrates towards the iron center. Already during the synthesis of the iron(II) precursor it became obvious, that small modifications of the ligand periphery change the properties of the resulting complexes significantly. Differences in the solubility of the complexes made the work up difficult and the reduction over sodium amalgam, well established in the case of the TIMEN^{mes} system, yielded unexpectedly two and threefold cyclo-metallated σ-aryl iron(II) and iron(III) complexes [110].

In Chap. 4, another TIMEN^R derivative is introduced. The new TIMEN^{3,5CF3} system combines features known from the TIMEN^{mes} ligand on one hand, and the TIMEN$^{tol/3,5xyl}$ derivatives on the other hand. It is capable of both nitride insertion and cyclometallation. Again, similar to the results and observations described in Chap. 3, the synthesis of an iron nitride complex supported by the new ligand could not be achieved. Instead, completely new reaction pathways were observed and investigated.

This work is in preparation for publication.

With our tripodal carbene ligand system TIMEN^{mes} we are able to synthesize an iron(IV) nitride complex that is remarkably air and moisture stable at RT. Accordingly, the complex shows low reactivity towards substrates. Upon oxidation the iron(IV) nitride complex undergoes an insertion reaction, but no iron(V) nitride complex could be isolated. The collaboration between Prof. Dr. Karsten Meyer (Friedrich-Alexander-University Erlangen-Nuremberg) and Prof. Dr. Jeremy Smith and his PhD student Jeremiah Scepaniak (New Mexico State University) opened the possibility for me to work with related tripodal carbene ligands PhB(^{R}Im)$_3^-$ (R = *tert*-butyl, mesityl), that feature a boron anchor and enable the synthesis of thermally stable iron(IV) nitride complexes. Within this collaboration, our research group was responsible for measurement and interpretation of Mößbauer, SQUID, and EPR data, for X-ray crystal structure determination as well as DFT studies.

The dissertation author was the secondary and tertiary researcher and author in Chaps. 5 and 6.

Chapter 5 describes the synthesis, structure, and reactivity of an iron(V) nitride. Jeremiah Scepaniak discovered that the reddish iron(IV) nitride precursor [PhB(^{tBu}Im)$_3$Fe^{IV}≡N] is readily oxidized by [Fe(Cp)$_2$]BAr$_{F24}$ at low temperatures in solution, thus generating the dark purple iron(V) nitride complex [PhB(^{tBu}Im)$_3$Fe^V≡N]BAr$_{F24}$ in high yield. In order to study its electronic and molecular structure, Mößbauer, SQUID, and EPR data are required, as well as an X-ray crystal structure determination. Due to the temperature sensitivity of the iron(V) nitride complex, Jeremiah Scepaniak provided the thermal stable precursors [PhB(^{tBu}Im)$_3$Fe^{IV}≡N] and [Fe(Cp)$_2$]BAr$_{F24}$ and I was in charge to reproduce the iron(V) nitride complex for analysis. By doing so I was able to obtain single crystals suitable for an X-ray crystal structure determination, which was carried out in cooperation with Dr. Frank W. Heinemann. For the investigation of the electronic structure I performed temperature-dependent ^{57}Fe Mößbauer measurements in cooperation with Dr. Jörg Sutter, together with X-band EPR measurements in cooperation with Dr. Marat M. Khusniyarov, who did DFT calculations as well [111].

Chapter 6 reports the first example of a four-coordinate iron(II) complex, [PhB(^{Mes}Im)$_3$Fe–N=PPh$_3$], that undergoes a thermally induced spin crossover. Jeremiah Scepaniak provided samples of this compound while I was able to obtain suitable single crystals and to perform variable temperature single-crystal X-ray diffraction experiments. The four-coordinate iron(II) complex, undergoes an $S = 0 \rightarrow S = 2$ spin transition with $T_C = 81$ K, as determined by variable temperature magnetic measurements, performed by me, while the spectra were interpreted and simulated with the help of Dr. Thomas D. Harris (University of California, Berkeley). Variable temperature Mößbauer spectroscopy was carried out by me with the help of Dr. Jörg Sutter. The structural changes have been interpreted in terms of electronic structure theory by Dr. Marat M. Khusniyarov [112].

References

1. W.v.E. Doering, A.K. Hoffmann, J. Am. Chem. Soc. **76**, 6162 (1954)
2. E.O. Fischer, A. Maasboel, Angew. Chem. **76**, 645 (1964)
3. G. Frenking, N. Fröhlich, Chem. Rev. **100**, 717 (2000)
4. W. Kirmse, *Carbene, Carbenoide und Carbenanaloge* (Verlag Chemie, Weinheim, 1969)
5. W.A. Herrmann, *Applied Homogeneous Catalysis with Organometallic Compounds*, vol. 3, 2nd edn. (VCH, Weinheim, 2002), p. 1078
6. F.E. Hahn, M.C. Jahnke, Angew. Chem. Int. Ed. **47**, 3122 (2008)
7. G. Bertrand, *Carbene Chemistry: From Fleeting Intermediates to Powerful reagents* (Fontis Media S. A. and Marcel Dekker, Lausanne, 2002)
8. X. Hu, *Metal Complexes of Tripodal N-Heterocyclic Carbene Ligands: Synthesis, Structure, Bonding, and Reactivity*. PhD thesis, University of California, San Diego (2004)
9. A.J. Arduengo, R. Krafczyk, Chemie in unserer Zeit **32**, 6 (1998)
10. W.W. Schoeller, J. Chem. Soc. Chem. Commun. **124**, (1980)
11. D. Bourissou, O. Guerret, F.P. Gabbai, G. Bertrand, Chem. Rev. **100**, 39 (2000)
12. G.B. Schuster, Adv. Phys. Org. Chem. **22**, 311 (1986)
13. R. Hoffmann, G.D. Zeiss, G.W. Van Dine, J. Am. Chem. Soc. **90**, 1485 (1968)
14. C.W. Bauschlicher Jr., H.F. Schaefer III, P.S. Bagus, J. Am. Chem. Soc. **99**, 7106 (1977)
15. D. Feller, W.T. Borden, E.R. Davidson, Chem. Phys. Lett. **71**, 22 (1980)
16. J.F. Harrison, J. Am. Chem. Soc. **93**, 4112 (1971)
17. J.F. Harrison, R.C. Liedtke, J.F. Liebman, J. Am. Chem. Soc. **101**, 7162 (1979)
18. C.D. Abernethy, G.M. Codd, M.D. Spicer, M.K. Taylor, J. Am. Chem. Soc. **125**, 1128 (2003)
19. J.B. Dumas, E.M. Péligot, Annales de chimie et de physique **58**, 5 (1835)
20. A. Geuther, Annalen der Chemie und Pharmacie **123**, 121 (1862)
21. H. Scheibler, Chem. Ber. **59**, 1022 (1926)
22. M. Schmeisser, H. Schroeter, H. Schilder, J. Massonne, F. Rosskopf, Chem. Ber. **95**, 1648 (1962)
23. M. Schmeisser, H. Schroter, Angew. Chem. **72**, 349 (1960)
24. H.W. Wanzlick, E. Schikora, Angew. Chem. **72**, 494 (1960)
25. H.J. Schoenherr, H.W. Wanzlick, Justus Liebigs Annalen der Chemie **731**, 176 (1970)

References 15

26. A.J. Arduengo III, R.L. Harlow, M. Kline, J. Am. Chem. Soc. **113**, 361 (1991)
27. T. Weskamp, V.P.W. Böhm, W.A. Herrmann, J. Organomet. Chem. **600**, 12 (2000)
28. C.M. Crudden, D.P. Allen, Coord. Chem. Rev. **248**, 2247 (2004)
29. S.P. Nolan (ed.), *N-Heterocyclic Carbenes in Synthesis* (Wiley-VCH, Weinheim, 2006)
30. F. Glorius, *N-Heterocyclic Carbenes in Transition Metal Catalysis*, vol. 21 (Springer, Berlin, 2007), p. 1
31. A.C. Hillier, G.A. Grasa, M.S. Viciu, H.M. Lee, C. Yang, S.P. Nolan, J. Organomet. Chem. **653**, 69 (2002)
32. E.A.B. Kantchev, C.J. O'Brien, M.G. Organ, Angew. Chem. Int. Ed. **46**, 2768 (2007)
33. A. Fürstner, Angew. Chem. Int. Ed. **39**, 3012 (2000)
34. S.J. Connon, S. Blechert, Angew. Chem. Int. Ed. **42**, 1900 (2003)
35. H. Clavier, K. Grela, A. Kirschning, M. Mauduit, S.P. Nolan, Angew. Chem. Int. Ed. **46**, 6786 (2007)
36. J.K. Huang, E.D. Stevens, S.P. Nolan, J.L. Petersen, J. Am. Chem. Soc. **121**, 2674 (1999)
37. W.A. Herrmann, Angew. Chem. Int. Ed. **41**, 1290 (2002)
38. W.A. Herrmann, M. Elison, J. Fischer, C. Kocher, G.R.J. Artus, Angew. Chem. Int. Ed. **34**, 2371 (1995)
39. W.A. Herrmann, C. Kocher, Angew. Chem. Int. Ed. **36**, 2163 (1997)
40. W.A. Herrmann, Angew. Chem. Int. Ed. **41**, 1290 (2002)
41. L. Sacconi, I. Bertini, J. Am. Chem. Soc. **90**, 5443 (1968)
42. X.L. Hu, I. Castro-Rodriguez, K. Olsen, K. Meyer, Organometallics **23**, 755 (2004)
43. D.M. Khramov, V.M. Lynch, C.W. Bielawski, Organometallics **26**, 6042 (2007)
44. L. Mercs, G.l. Labat, A. Neels, A. Ehlers, M. Albrecht, Organometallics **25**, 5648 (2006)
45. M.D. Sanderson, J.W. Kamplain, C.W. Bielawski, J. Am. Chem. Soc. **128**, 16514 (2006)
46. H. Jacobsen, A. Correa, C. Costabile, L. Cavallo, J. Organomet. Chem. **691**, 4350 (2006)
47. D. Nemcsok, K. Wichmann, G. Frenking, Organometallics **23**, 3640 (2004)
48. R. Tonner, G. Heydenrych, G. Frenking, Chem Asian J **2**, 1555 (2007)
49. N.M. Scott, R. Dorta, E.D. Stevens, A. Correa, L. Cavallo, S.P. Nolan, J. Am. Chem. Soc. **127**, 3516 (2005)
50. H. Jacobsen, A. Correa, A. Poater, C. Costabile, L. Cavallo, Coord. Chem. Rev. **253**, 687 (2009)
51. J. Chatt, L.A. Duncanson, J. Chem. Soc. 2939 (1953)
52. R.R. Schrock, Acc. Chem. Res. **30**, 9 (1997)
53. T.A. Betley, J.C. Peters, Inorg. Chem. **42**, 5074 (2003)
54. H.A. Mayer, W.C. Kaska, Chem. Rev. **94**, 1239 (1994)
55. S. Trofimenko, J. Chem. Educ. **82**, 1715 (2005)
56. N. Burzlaff, Angew. Chem. Int. Ed. **48**, 5580 (2009)
57. S. Trofimenko, J. Am. Chem. Soc. **88**, 1842 (1966)
58. S. Trofimenko, Prog. Inorg. Chem. **34**, 115 (1986)
59. S. Trofimenko, Chem. Rev. **93**, 943 (1993)
60. N. Kitajima, Y. Moro-oka, Chem. Rev. **94**, 737 (1994)
61. D.L. Reger, C.A. Little, A.L. Rheingold, M. Lam, L.M. Liable-Sands, B. Rhagitan, T. Concolino, A. Mohan, G.J. Long, V. Briois, F. Grandjean, Inorg. Chem. **40**, 1508 (2001)
62. D.L. Reger, C.A. Little, A.L. Rheingold, R. Sommer, G.J. Long, Inorg. Chim. Acta **316**, 65 (2001)
63. H. Pfeiffer, A. Rojas, J. Niesel, U. Schatzschneider, Dalton Trans. (22), 4292 (2009)
64. R.R. Schrock, D.V. Yandulov, Science **301**, 76 (2003)
65. R.A. Kinney, D.G.H. Hetterscheid, B.S. Hanna, R.R. Schrock, B.M. Hoffman, Inorg. Chem. **49**, 704 (2009)
66. M.R. Reithofer, R.R. Schrock, P. Müller, J. Am. Chem. Soc. **132**, 8349 (2010)
67. D.C. Lacy, R. Gupta, K.L. Stone, J. Greaves, J.W. Ziller, M.P. Hendrich, A.S. Borovik, J. Am. Chem. Soc. **132**, 12188 (2010)
68. C.E. MacBeth, A.P. Golombek, V.G. Young Jr., C. Yang, K. Kuczera, M.P. Hendrich, A.S. Borovik, Science **289**, 938 (2000)
69. D.M. Jenkins, T.A. Betley, J.C. Peters, J. Am. Chem. Soc. **124**, 11238 (2002)

16   1 Introduction to Tripodal N-Heterocyclic Carbene Iron Complexes

70. T.A. Betley, J.C. Peters, J. Am. Chem. Soc. **126**, 6252 (2004)
71. S.D. Brown, T.A. Betley, J.C. Peters, J. Am. Chem. Soc. **125**, 322 (2003)
72. M.P. Hendrich, W. Gunderson, R.K. Behan, M.T. Green, M.P. Mehn, T.A. Betley, C.C. Lu, J.C. Peters, Proc. Natl. Acad. Sci. **103**, 17107 (2006)
73. H.V.R. Dias, W.C. Jin, Tetrahedron Lett. **35**, 1365 (1994)
74. H. Nakai, Y. Tang, P. Gantzel, K. Meyer, Chem. Commun. (1), 24 (2003)
75. U. Kernbach, M. Ramm, P. Luger, W.P. Fehlhammer, Angew. Chem. Int. Ed. **35**, 310 (1996)
76. R. Fränkel, C. Birg, U. Kernbach, T. Habereder, H. Nöth, W.P. Fehlhammer, Angew. Chem. Int. Ed. **40**, 1907 (2001)
77. R. Fränkel, U. Kernbach, M. Bakola-Christianopoulou, U. Plaia, M. Suter, W. Ponikwar, H. Nöth, C. Moinet, W.P. Fehlhammer, J. Organomet. Chem. **617–618**, 530 (2001)
78. J.J. Scepaniak, J.A. Young, R.P. Bontchev, J.M. Smith, Angew. Chem. Int. Ed. **48**, 3158 (2009)
79. J.J. Scepaniak, M.D. Fulton, R.P. Bontchev, E.N. Duesler, M.L. Kirk, J.M. Smith, J. Am. Chem. Soc. **130**, 10515 (2008)
80. K. Chen, M. Costas, J. Kim, A.K. Tipton, L. Que, J. Am. Chem. Soc. **124**, 3026 (2002)
81. M. Costas, M.P. Mehn, M.P. Jensen, L. Que Jr., Chem. Rev. **104**, 939 (2004)
82. J.-U. Rohde, J.-H. In, M.H. Lim, W.W. Brennessel, M.R. Bukowski, A. Stubna, E. Münck, W. Nam, L. Que, Science **299**, 1037 (2003)
83. K. Meyer, E. Bill, B. Mienert, T. Weyhermüller, K. Wieghardt, J. Am. Chem. Soc. **121**, 4859 (1999)
84. M. Aliaga-Alcalde, S.D. George, B. Mienert, E. Bill, K. Wieghardt, F. Neese, Angew. Chem. Int. Ed. **44**, 2908 (2005)
85. C.A. Grapperhaus, B. Mienert, E. Bill, T. Weyhermüller, K. Wieghardt, Inorg. Chem. **39**, 5306 (2000)
86. J.F. Berry, E. Bill, E. Bothe, S.D. George, B. Mienert, F. Neese, K. Wieghardt, Science **312**, 1937 (2006)
87. G. Ertl, Chem. Rec. **1**, 33 (2001)
88. B.M. Hoffman, D.R. Dean, L.C. Seefeldt, Acc. Chem. Res. **42**, 609 (2009)
89. G. Ertl, Angew. Chem. Int. Ed. **29**, 1219 (1990)
90. O. Einsle, F.A. Tezcan, S.L.A. Andrade, B. Schmid, M. Yoshida, J.B. Howard, D.C. Rees, Science **297**, 1696 (2002)
91. D. Lukoyanov, V. Pelmenschikov, N. Maeser, M. Laryukhin, T.-C. Yang, L. Noodleman, D.R. Case, D.A. Case, L.C. Seefeldt, B.M. Hoffman, Inorg. Chem. **46**, 11437 (2007)
92. T.-C. Yang, N.K. Maeser, M. Laryukhin, H.-I. Lee, D.R. Dean, L.C. Seefeldt, B.M. Hoffman, J. Am. Chem. Soc. **127**, 12804 (2005)
93. D.S. Scholl, J.A. Steckel, *Density Functional Theory: A Practical Introduction* (Wiley, Hoboken, 2009)
94. K. Meyer, J. Bendix, N. Metzler-Nolte, T. Weyhermüller, K. Wieghardt, J. Am. Chem. Soc. **120**, 7260 (1998)
95. C.J. Ballhausen, H.B. Gray, Inorg. Chem. **1**, 111 (1962)
96. J.R. Winkler, H.B. Gray, J. Am. Chem. Soc. **105**, 1373 (1983)
97. J.R. Winkler, H.B. Gray, Inorg. Chem. **24**, 346 (1985)
98. J. Bendix, J. Am. Chem. Soc. **125**, 13348 (2003)
99. K. Meyer, J. Bendix, E. Bill, T. Weyhermüller, K. Wieghardt, Inorg. Chem. **37**, 5180 (1998)
100. A. Niemann, U. Bossek, G. Haselhorst, K. Wieghardt, B. Nuber, Inorg. Chem. **35**, 906 (1996)
101. J. Bendix, K. Meyer, T. Weyhermüller, E. Bill, N. Metzler-Nolte, K. Wieghardt, Inorg. Chem. **37**, 1767 (1998)
102. J. Bendix, R.J. Deeth, T. Weyhermüller, E. Bill, K. Wieghardt, Inorg. Chem. **39**, 930 (2000)
103. M. Aliaga-Alcalde, S.D. George, B. Mienert, E. Bill, K. Wieghardt, F. Neese, Angew. Chem. Int. Ed. **44**, 2908 (2005)
104. T.A. Betley, J.C. Peters, J. Am. Chem. Soc. **125**, 10782 (2003)
105. D.M. Jenkins, A.J. Di Bilio, M.J. Allen, T.A. Betley, J.C. Peters, J. Am. Chem. Soc. **124**, 15336 (2002)
106. C.M. Thomas, N.P. Mankad, J.C. Peters, J. Am. Chem. Soc. **128**, 4956 (2006)

References

107. C.M. Thomas, J.C. Peters, Angew. Chem. Int. Ed. **45**, 776 (2006)
108. X. Hu, K. Meyer, J. Am. Chem. Soc. **126**, 16322 (2004)
109. C.S. Vogel, F.W. Heinemann, J. Sutter, C. Anton, K. Meyer, An iron nitride complex. Angew. Chem. Int. Ed. **47**, 2681–2684 (2008)
110. C.S. Vogel, F.W. Heinemann, M.M. Khusniyarov, K. Meyer, Unexpected reactivity resulting from modifications of the ligand periphery: synthesis, structure, and spectroscopic properties of iron complexes of new tripodal N-heterocyclic carbene (NHC) ligands: invited article dedicated to Prof. Arnold Rheingold. Inorg. Chim. Acta **364**, 226–236 (2010)
111. J.J. Scepaniak, C.S. Vogel, M.M. Khusniyarov, F.W. Heinemann, K. Meyer, J.M. Smith, Synthesis, structure, and reactivity of an iron(V) nitride. Science **331**, 1049–1052 (2011)
112. J.J. Scepaniak, T.D. Harris, C.S. Vogel, J. Sutter, K. Meyer, J.M. Smith, Spin crossover in a four-coordinate iron(II) complex. J. Am. Chem. Soc. **133**, 3824–3827 (2011)

Chapter 2
TIMEN^{mes}: An Iron Nitride Complex

2.1 Introduction

Coordination compounds of iron in high oxidation states have been invoked as reactive intermediates in biocatalyses. Iron(IV) ferryl species are examples of highly reactive compounds that have long been known to be at the catalytic centers of oxygenases [1]. Supported by X-ray diffraction studies on nitrogenase, the iron nitride moiety has recently been suggested to be present at the site of biological nitrogen reduction [2]. As a result, well-characterized high-valent iron complexes have been sought as biomimetic models for transformations mediated by iron-containing enzymes. To gain understanding of iron nitride reactivity and the possible role of such species in biocatalysis, insight into the molecular and electronic structure of complexes stabilizing the [FeN] synthon is highly desirable. Whereas significant progress has been made in the synthesis and spectroscopic elucidation of Fe=NR and Fe$\equiv$N species [3–10], X-ray crystallographic characterization of a complex with a terminal Fe$\equiv$N functionality has not been accomplished [11, 12]. The first mononuclear Fe(IV)=O entity crystallographically characterized was stabilized in an octahedral environment provided by a macrocyclic *tetra*-N-methylated cyclam ligand [13]. Similar cyclam derivatives also allow the stabilization and detailed spectroscopic characterization of octahedral Fe(V) and Fe(VI) nitride complexes in unusually high oxidation states [4, 5, 8]. Recently, Peters and Betley developed a stunningly redox-rich iron system employing the tripodal *tris*(phosphino)borate ligand system (PhB^{R}P$_3^-$), which stabilizes tetrahedral L$_3$Fe=N$_x$ species in oxidation states ranging from +I to +IV [11]. Remarkably, this ligand system enabled the first room-temperature spectroscopic characterization of a terminal Fe(IV) nitride species. Concentration-dependent coupling to the Fe(I)–N$_2$–Fe(I) dinuclear product, however, prevents crystallization of this nitride species.

C. S. Vogel, *High- and Low-Valent tris-N-Heterocyclic Carbene Iron Complexes*,
Springer Theses, DOI: 10.1007/978-3-642-27254-7_2,
© Springer-Verlag Berlin Heidelberg 2012

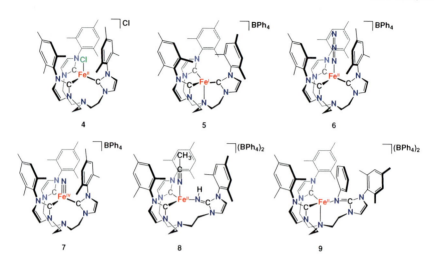

Chart 2.1 Overview of the iron complexes derived from the tripodal TIMEN^{mes} ligand

Inspired by this work, we sought to prepare an iron nitride by employing the sterically more encumbering N-anchored *tris*(carbene) ligand, *tris*[2-(3-mesityl-*im*idazol-2-ylidene)*e*thyl]amine (TIMEN^{mes}, mesityl (mes)) [14].

This chapter deals with the synthesis, spectroscopy, and most significantly, the X-ray diffraction analysis of a discrete iron nitride complex. Chart 2.1 gives an overview of the series of iron complexes bearing the TIMEN^{mes} ligand, which were synthesized and characterized in this context.

2.2 Results and Discussion

2.2.1 General Synthetic Procedure for the Precursors of the TIMEN^R System

Only a few N-substituted imidazoles such as the N-methyl and N-benzyl imidazoles are commercially available. For custom-tailored ligand environments with varying steric demand, it is therefore inevitable to synthesize the desired imidazoles by one-pot condensation reactions of aryl amine, formaldehyde, glyoxal, and ammonium acetate [15]. In the course of this work, the imidazolium precursors, *tris*-[2-(3-arylimidazolium)ethyl]amine [H$_3$TIMEN^R]$^{3+}$ (R = mesityl (mes), tolyl (tol), 3,5-xylyl (3,5xyl) 3,5-trifluoromethylphenyl (3,5CF3)) were prepared by quaternization of the functionalized N-arylimidazoles with *tris*(chloroethyl)amine (Scheme 2.1) [16]. Chart 2.2 gives an overview of all TIMEN^R ligands available to date.

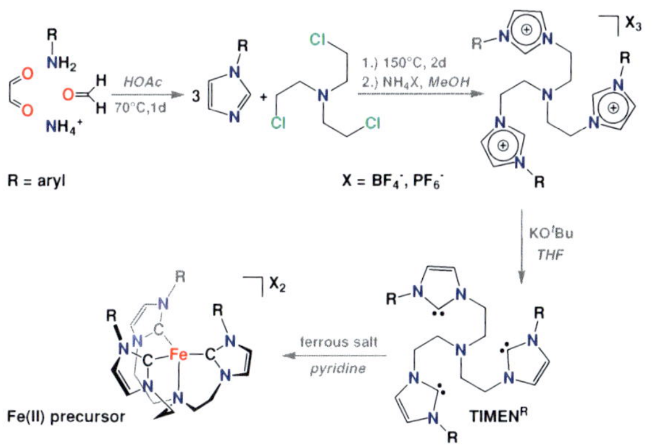

2.2 Results and Discussion

Scheme 2.1 Synthesis of TIMEN^R and the Fe(II) precursor complexes

Treatment of $[H_3TIMEN^R](Cl)_3$ in methanol with ammonium hexafluorophosphate results in complete substitution of chloride and formation of $[H_3TIMEN^R](PF_6)_3$. Deprotonation of the imidazolium precursors with a strong base, like potassium *tert*-butoxide, yields the free *tris*(carbene) ligands TIMEN^R (*tris*-[2-(3-*aryl*-**im**idazol-2-ylidene)-**e**thyl]-ami**n**e), which can be isolated and stored under nitrogen. However, it is difficult to remove the deprotonation byproduct *tert*-butanol. The free carbenes are highly soluble in Et$_2$O and THF and even slightly soluble in hexane and *n*-pentane. Thus, there is no possibility to remove the alcohol by washing and because of the high boiling point it is difficult to remove *tert*-butanol in vacuo. This is also the reason why there is no exact yield of the carbene specified in the experimental section. The free *tris*(carbene) ligands were characterized by ^{1}H- and ^{13}C-NMR spectroscopy. The characteristic ^{13}C-NMR signals for the deprotonated carbene carbon atoms can be found between 210 and 215 ppm. After deprotonation of the imidazolium salt, the free carbene can be treated with ferrous chloride which yields an iron(II) complex. Depending on the solvent used in the synthesis, the remaining *tert*-butanol can cause further problems. Using pyridine, which acts as a mild base, coordination of the iron ion by the carbene ligand occurs smoothly. By changing the solvent from pyridine to Et$_2$O and THF, the mixture of *tert*-butanol and ferrous chloride creates acidic conditions, which result in protonation of the carbene. This back reaction reforms the imidazolium precursor faster than the free carbene is able to chelate to the iron center.

2.2.2 The TIMEN^{mes} Ligand and its Coordination to Iron: Synthesis Towards an Elusive Iron Nitride Complex

Under inert atmosphere, treatment of TIMEN^{mes} (2.3) with one equivalent of anhydrous ferrous chloride in pyridine at room temperature yields the four-coordinate Fe(II) complex [(TIMEN^{mes})Fe(Cl)]Cl (2.4) as analytically pure,

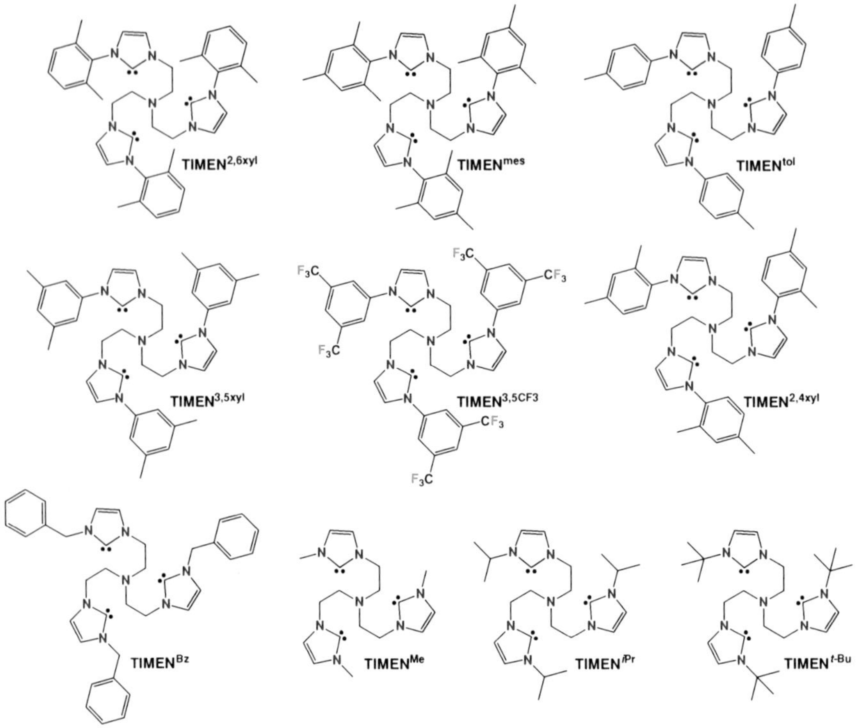

Chart 2.2 Overview of the tripodal TIMEN^R ligand family

white powder in 80% yield (Scheme 2.1). The ^{1}H-NMR spectrum of **4** in solution is consistent with a paramagnetic high-spin iron(II) complex. It shows eleven paramagnetically shifted resonances in the range between -6 and 78 ppm. Only three signals at 5.14, -0.32 and -5.90 ppm, which belong to the methyl groups at the aryl rings, can be clearly assigned due to peak integration (Fig. 2.1).

The zero-field Mößbauer spectrum of **4** is shown in Fig. 2.2. This spectrum recorded from a crystalline sample of **4** at 77 K displays a quadrupole doublet with an isomer shift $\delta = 0.71(1)$ mm s^{-1}, a quadrupole splitting parameter $\Delta E_Q = 1.78(1)$ mm s^{-1}, and a line width $\Gamma_{FWHM} = 0.32(1)$ mm s^{-1}. These parameters are characteristic of a four-coordinate d^6 high-spin ($S = 2$) Fe(II) complex [17]. The SQUID magnetization measurements of **4** also confirm an $S = 2$ ground state for this high-spin Fe(II) complex (see Fig. 2.3).

The molecular structure shows the iron center to coordinate in a distorted trigonal pyramidal fashion. A long Fe1$\cdots$N1 distance of 3.008(3) Å indicates that the nitrogen anchor is non-coordinating, while the three carbene carbon atoms bind with a mean bond distance of 2.135 Å. The iron center is 0.439(2) Å above the trigonal plane, defined by the three carbene carbon atoms. For the deviation of the iron center from the carbene plane the term out-of-plane (oop) shift, $d_{Fe\ oop}$ has

2.2 Results and Discussion

Fig. 2.1 ^{1}H-NMR spectrum of [(TIMEN^{mes})Fe(Cl)]Cl (**4**) (recorded in DMSO-d_6 at RT)

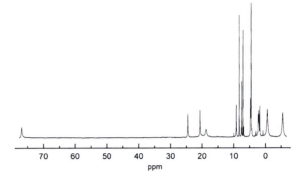

Fig. 2.2 Zero-field Mößbauer spectrum of [(TIMEN^{mes})Fe(Cl)]Cl (**4**) (recorded in solid state at 77 K; $\delta = 0.71(1)$ mm s^{-1}, $\Delta E_Q = 1.78(1)$ mm s^{-1}, $\Gamma_{FWHM} = 0.32(1)$ mm s^{-1})

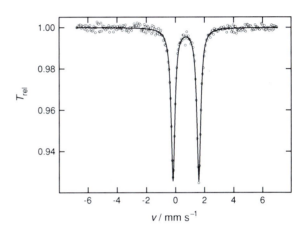

Fig. 2.3 Plot of the effective magnetic moment, μ_{eff}, versus temperature from SQUID magnetization measurements for two independently prepared samples of [(TIMEN^{mes})Fe(Cl)]Cl (**4**) ($\mu_{eff} = 4.89$ μ_B at 300 K, simulated parameters: $g = 2.0$, $|D| = 4.2$ cm^{-1}, diamagnetic impurity = 1.0%)

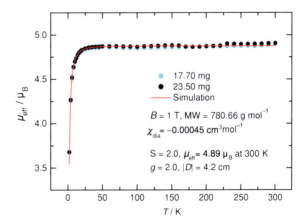

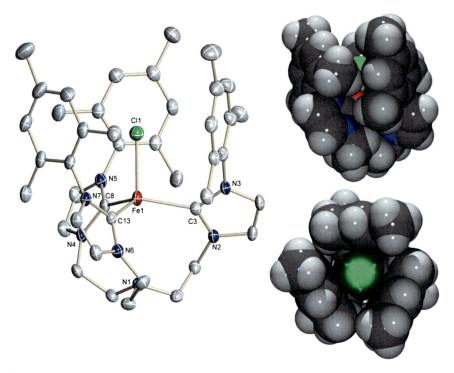

Fig. 2.4 Molecular structure of the complex cation of [(TIMEN^{mes})Fe(Cl)]Cl (**4**) in crystals of **4** · 3 MeCN (50% probability ellipsoids). Hydrogen atoms, Cl$^-$ anion, and co-crystallized solvents are omitted for clarity. Space-filling representations of the cation **4** *side view* (*right, top*) and *top view* (*right, bottom*). Selected bond distances (Å): Fe1–Cl1 2.2768(8), Fe1–C3 2.128(3), Fe1–C8 2.138(3), Fe1–C13 2.138(2), Fe1⋯N1 3.008(3), $d_{Fe\ oop}$ 0.439(2)

been established. The axial chloro ligand occupies the protective cavity formed by the three mesityl substituents (for principle bond distances see Fig. 2.4).

Reduction of the Fe(II) precursor complex **4** over sodium amalgam in a solution of NaBPh$_4$ in THF leads to the formation of deep red solutions of the reduced compound. The reaction mixture was stirred over night, the solution was filtered, concentrated, and cooled to -35 °C to yield dark red crystals of the Fe(I) complex [(TIMEN^{mes})Fe]BPh$_4$ (**5**, Scheme 2.2). Compound **5** is NMR silent but EPR active. An axial EPR spectrum with *g*-values of 4.20 and 2.025 confirms the $S = 3/2$ ground state of this molecule (Fig. 2.5).

The zero-field Mößbauer spectrum of **5** is shown in Fig. 2.6. The spectrum of a crystalline sample of **5** at 77 K displays a quadrupole doublet with an isomer shift $\delta = 0.64(1)$ mm s^{-1}, a quadrupole splitting parameter $\Delta E_Q = 2.34(1)$ mm s^{-1}, and a line width $\Gamma_{FWHM} = 0.29(1)$ mm s^{-1}. The results of a single crystal X-ray analysis show that the molecular structure of the core-complex of the [TIMEN^{mes}Fe]$^+$ system is barely changed upon reduction (Fig. 2.7).

2.2 Results and Discussion

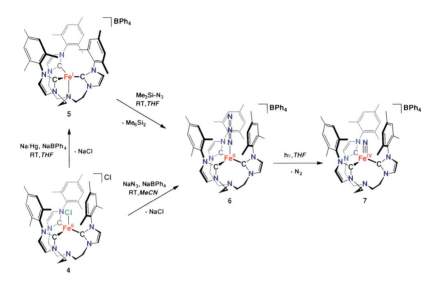

Scheme 2.2 Synthetic routes towards the iron(IV) nitride complex [(TIMEN^{mes})Fe(N)]BPh$_4$ (**7**)

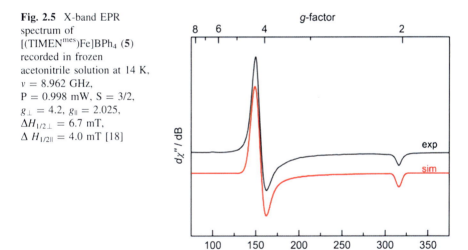

Fig. 2.5 X-band EPR spectrum of [(TIMEN^{mes})Fe]BPh$_4$ (**5**) recorded in frozen acetonitrile solution at 14 K, v = 8.962 GHz, P = 0.998 mW, S = 3/2, $g_\perp$ = 4.2, $g_\parallel$ = 2.025, $\Delta H_{1/2\perp}$ = 6.7 mT, $\Delta H_{1/2\parallel}$ = 4.0 mT [18]

However, upon chloride removal the nitrogen anchor binds to the iron center. This is indicated by the shortened Fe1–N1 bond distance of 2.275(2) Å. Also interesting to mention is the observation that the mean value of the iron carbene distances shortens significantly upon reduction from 2.135(3) Å in the Fe(II) complex **5** to 2.023(2) Å in the Fe(I) species **4**. This can be accounted to strong π-backbonding within the iron carbene bond. A similar trend is observed in the

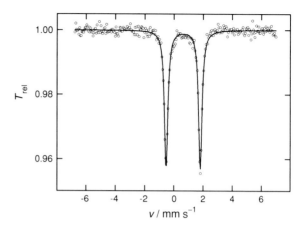

Fig. 2.6 Zero-field Mößbauer spectrum of [(TIMEN^{mes})Fe]BPh$_4$ (5) (recorded in solid state at 77 K; $\delta = 0.64(1)$ mm s^{-1}, $\Delta E_Q = 2.34(1)$ mm s^{-1}, $\Gamma_{FWHM} = 0.29(1)$ mm s^{-1})

Mößbauer spectrum, where the isomer shift changes in an opposite direction to what is expected. The isomer shift allows conclusions on possible oxidation states of the examined compounds. It is generally applicable that with increasing electron density at the ^{57}Fe nucleus the isomer shift, δ, becomes more negative. Going from elemental iron with an [Ar]3$d^6$4s^2 electron configuration to Fe(IV), first of all the 4s electrons are removed. Fe(II) compounds with an [Ar]3d^6 electron configuration have no 4s electrons left. Upon further oxidation the 3d electrons, which predominantly influence the electron density at the nucleus, are removed. The closer the outer electron can get to the nucleus, the more it can penetrate the density distribution of the inner shell electrons and the atomic nucleus. For many purposes, atomic nuclei can be treated as mathematical point charges. But when electrons penetrate the nuclear volume the shape and size of the nucleus does play a role. The usual "far-field" approximation is then invalid and these "near-field" effects lead to small corrections to all terms in the multipole expansion for the electrostatic interaction between nuclei and electrons. The correction to the monopole term corresponds to experimentally well-known phenomena: the isotope shift in atomic spectroscopy and of particular interest here, the isomer shift in Mößbauer spectroscopy [19]. Because of the possibility that electrons, mainly s electrons, can penetrate into the nucleus one could speak of an electron density at the nucleus, expressed by the square wave function at the position $r = 0$, $|\psi(0)|^2$. The electrostatic interaction between nucleus and electron (electric monopole interaction) changes the energy level of the nucleus. The volume of the nucleus in the excited state is different from that in the ground state, and therefore, also the strength of the electric monopole interaction differs in both states. The consequence is a different shift of the energy level of the ground state and the exited state. The virtual transition energy E_o (the required energy for the nuclear transition between the ground state and the excited state, without any perturbation and interaction), is then exchanged by the real transition energies E_S (emitting nucleus, excited state; radiation source) and E_A (absorber nucleus, ground state; sample).

2.2 Results and Discussion

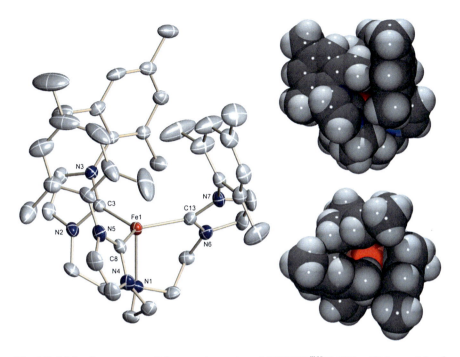

Fig. 2.7 Molecular structure of the complex cation of [(TIMEN^{mes})Fe]BPh$_4$ (**5**) in crystals of **5** · 2 THF · 0.5 Et$_2$O (50% probability ellipsoids). Hydrogen atoms, BPh$_4^-$ anion, and co-crystallized solvents are omitted for clarity. Space-filling representations of the cation **5** *side view* (*right, top*) and *top view* (*right, bottom*). Selected bond distances (Å): Fe1–C3 2.032(2), Fe1–C8 2.024(2), Fe1–C13 2.012(2), Fe1–N1 2.275(2), $d_{\text{Fe oop}}$ 0.191(2)

The differences

$$\Delta E(S) = E_S - E_0 = \text{const} \cdot |\psi(0)|_E^2 \cdot \left(R_e^2 - R_g^2\right) \quad (2.1)$$

$$\text{and } \Delta E(A) = E_A - E_0 = \text{const} \cdot |\psi(0)|_A^2 \cdot \left(R_e^2 - R_g^2\right) \quad (2.2)$$

are determined by two factors, the electron density at the nucleus and the different spatial dimensions of the nucleus in the excited and the ground state (expressed by the difference of the mean square nuclear radii: $R_e^2 - R_g^2$). If the environment of the Mößbauer nucleus differs in the radiation source and the absorber, the electron density at the nuclei are different ($|\psi(0)|_E^2 \neq |\psi(0)|_e^2$); thus the energy differences $\Delta E(S)$ and $\Delta E(A)$ are unequal aswell. The difference is the isomer shift, δ, which can be measured.

$$\delta = \Delta E(A) - \Delta E(S) = E_A - E_S \quad (2.3)$$

$$\delta = \text{const} \cdot \left\{|\psi(0)|_A^2 - \psi(0)|_S^2\right\} \cdot \left(R_e^2 - R_g^2\right) \quad (2.4)$$

The factor $(R_e^2 - R_g^2)$ in (2.4) of the isomer shift is a negative term in case of the Mößbauer nucleus ^{57}Fe. Accordingly, the shift values for Fe(IV) species with high oxidation state are at the negative end of the scale, whereas low-valent iron compounds have more positive isomer shifts [20, 21]. As a result, it can be expected when going from Fe(II) to Fe(I) the isomer shift should increase with decreasing oxidation state; however, the reverse is observed. Again, π-backbonding within the iron carbene bond causes this effect. The more electron density is withdrawn by π-backbonding out of the $3d$ orbitals the less is the shielding of the nucleus. This explains the more negative isomer shift of the Fe(I) species **5** compared to its Fe(II) counterpart **4**.

The Fe(I) compound **5** is highly reactive towards oxygen and moisture. Complex **5** can easily be oxidized, even by chlorinated solvents like chloroform and methylene chloride. Addition of excess trimethylsilyl azide to the Fe(I) complex **5** in THF results in elimination of Me_6Si_2 and formation of a light yellow solution, from which pale yellow crystals of divalent $[(TIMEN^{mes})Fe(N_3)]BPh_4$ (**6**, Scheme 2.2) can be isolated in 79% yield.

In search of an alternative synthetic route for the preparation of the ^{15}N labeled azide complex, it turned out that a salt metathesis reaction between the Fe(II) chloro precursor **4** and an excess of sodium azide also yields **6**. The azide complex is easily identified by the intense and very characteristic $\nu_{as}(N_3)$ IR vibrational band at 2,094 cm^{-1}. The crystal structure of compound **6** (see Fig. 2.8) features an unusual, close to linear azide coordination ($\angle(Fe-N_\alpha-N_\beta) = 174.5(2)°$ in **6**) at the Fe(II) *tris*(carbene) complex. Such linear azide coordination to a metal center is rare but has been observed before in sterically demanding ligand environments, in which similar cylindrical cavities are generated around the metal center [22]. The Fe–N_α azide distance in **6** is determined to be 1.947(2) Å. The amine anchor is not bound [$d(Fe\cdots N1) = 3.244(2)$ Å] leaving the Fe(II) ion in a trigonal-pyramidal coordination environment, in which the metal center is located 0.554(2) Å above the trigonal plane formed by the three carbene carbon donor atoms of the NHC ligand.

The azide compound **6** is light sensitive and readily releases dinitrogen to form the deeply purple colored product $[(TIMEN^{mes})Fe(N)]BPh_4$ (**7**, Scheme 2.2). Accordingly, exposure of a pale yellow solution of **6** in THF to the light of a Xenon arc lamp at room temperature results in gradual formation of a purple solution and N_2 gas evolution. While the electronic absorption spectrum of the colorless azide complex **6** is featureless in the visible region, the spectrum of the photolysis product **7** (Fig. 2.9) shows an intense absorption band centered at λ_{max} 520 nm ($\varepsilon = 1,980 \ M^{-1} \ cm^{-1}$) and given rise to the purple color.

Photolysis of the iron azide **6** was continued until no more gas evolution was observed and the azide vibrational band could no longer be observed by IR spectroscopy (see Fig. 2.10). Instead, in the spectrum of the purple photolysis product a new band appeared at 1,008 cm^{-1}, which is assigned to the iron nitride band $\nu \ (Fe \equiv {}^{14}N)$.

Samples of the ^{15}N-labeled isotopomer (50% ^{15}N-labeled at the terminal nitride) show an additional band at 982 cm^{-1} (Fig. 2.10). This 26 cm^{-1} shift of

2.2 Results and Discussion

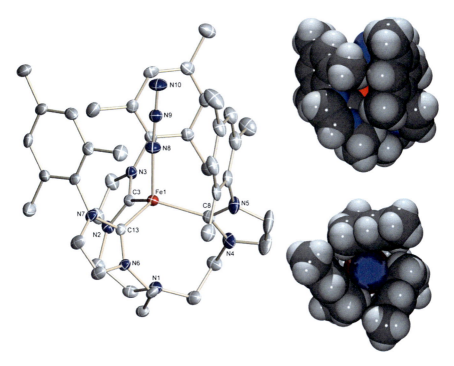

Fig. 2.8 Molecular structure of the complex cation of [(TIMEN^{mes})Fe(N$_3$)]BPh$_4$ (**6**) in crystals of **6** · 2.5 THF (50% probability ellipsoids). Hydrogen atoms, BPh$_4^-$ anion, and co-crystallized solvents are omitted for clarity. Space-filling representations of the cation **6** *side view* (*right, top*) and *top view* (*right, bottom*). Selected bond distances and angles (Å, °): Fe1–N8 1.947(2), Fe1–C3 2.104(2), Fe1–C8 2.110(3), Fe1–C13 2.111(2), Fe1⋯N1 3.244(2), $d_{Fe\ oop}$ 0.554(2), Fe1–N8–N9 174.5(2), N8–N9–N10 179.6(3)

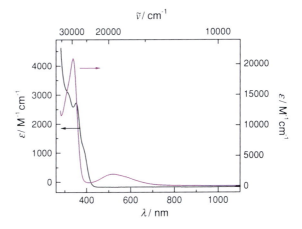

Fig. 2.9 Electronic absorption spectra of [(TIMEN^{mes})Fe(N$_3$)]BPh$_4$ (**6**) (black) and [(TIMEN^{mes})Fe(N)]BPh$_4$ (**7**) (magenta), recorded in THF

Fig. 2.10 *Top*: IR spectrum of [(TIMEN^mes)Fe(N)]BPh₄ (**7**) (KBr pellet); *bottom*: IR spectrum of [(TIMEN^mes)Fe(¹⁴N)]BPh₄ (*black*) and ¹⁵N-labeled isotopomer [(TIMEN^mes)Fe(¹⁵N)]BPh₄ (*red*, 50% ¹⁵N-labeled at the terminal nitride)

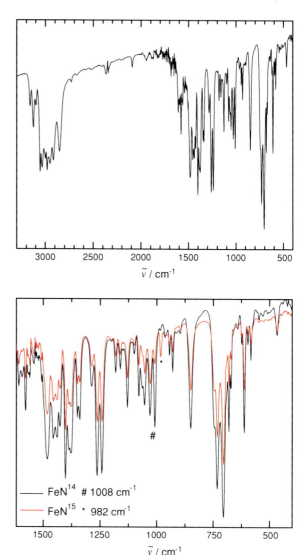

the v (Fe≡¹⁵N) stretch to lower frequency is predicted from the reduced mass calculation for a harmonic oscillator (calcd: 28 cm⁻¹). In the ¹H-NMR spectrum (Fig. 2.11) the purple, diamagnetic reaction product shows the expected resonances for the functionalized *tris*(carbene) chelator coordinated to the iron center of **7**. Additionally, the ¹⁵N-NMR spectrum of ¹⁵N-labeled **7** (Fig. 2.12) exhibits a resonance signal at $\delta = 1{,}121$ ppm (referenced to NH₃). Slow diffusion of diethyl ether into a THF solution of **7** yields purple crystals that are air stable at room temperature and suitable for an X-ray structure analysis.

2.2 Results and Discussion

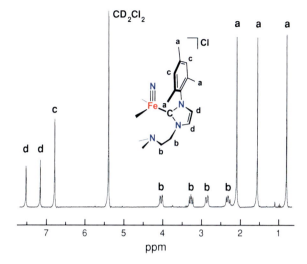

Fig. 2.11 ^{1}H-NMR spectrum of [(TIMEN^{mes})Fe(N)]BPh$_4$ (**7**) (recorded in DMSO-d_6 at RT)

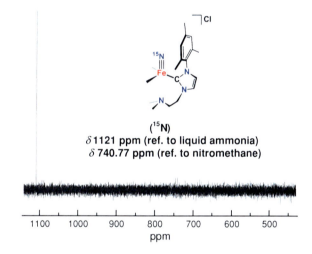

Fig. 2.12 ^{15}N-NMR spectrum of ^{15}N-labeled isotopomer [(TIMEN^{mes})Fe(N)]BPh$_4$ (**7**) (recorded in DMSO-d_6 at RT)

The molecular structure of [(TIMEN^{mes})Fe(N)]BPh$_4$ (Fig. 2.13) confirms the light-induced transformation of the axial azide to a terminal nitride ligand with a short Fe≡N bond of 1.526(2) Å. Such short M≡N distances are typically observed for transition-metal complexes with triply bonded nitride ligands (M=Cr, Mn) [8, 23–26]. The Fe≡N distance in **7** is also very similar to the XAS spectroscopically determined distances of the Fe(IV) complexes [(PhB^{R}P$_3$)Fe(N)] [d(Fe≡N) = 1.51–1.55(2) Å, R = i–Pr, CH$_2$-Cy] by Peters and co-workers, and of the octahedral Fe(VI) nitrido species [(Me$_3$cy-ac)Fe(N)] [d(Fe≡N) = 1.57 Å] reported by Wieghardt and co-workers [4]. As observed for the Fe(II) azide precursor **6**, the iron center in the nitrido complex **7** is coordinated in a trigonal-pyramidal fashion but with the Fe(IV) ion located at only 0.427(3) Å above the

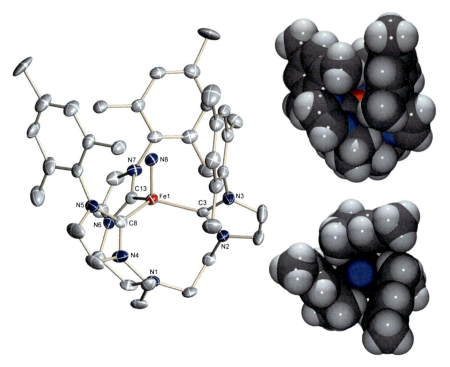

Fig. 2.13 Molecular structure of the complex cation of [(TIMEN^{mes})Fe(N)]BPh$_4$ (**7**) in crystals of **7** · Et$_2$O (50% probability ellipsoids). Hydrogen atoms, BPh$_4^-$ anion, and co-crystallized solvents are omitted for clarity. Space-filling representations of the cation **7** *side view* (*right, top*) and *top view* (*right, bottom*). Selected bond distances (Å): Fe1–N8 1.526(2), Fe1–C3 1.987(3), Fe1–C8 1.955(5), Fe1–C13 1.946(4), Fe1⋯N1 3.120(2), $d_{Fe\ oop}$ 0.427(3)

trigonal *tris*(carbene) plane [in **6**: 0.554(2) Å]. Evidently, this smaller displacement from the trigonal coordination plane in **7** is a consequence of increased interaction of the anchoring amine nitrogen [d(Fe1⋯N1) = 3.120(2) Å in **7** versus 3.244(2) Å in **6**] and the carbene ligand with the high-valent Fe(IV) ion. This stronger coordination of the carbene in **7** is also reflected by a significantly shorter average Fe–C distance of 1.941(5) Å compared to the weaker binding in the Fe(II) azide precursor **6** with a d(Fe–C)$_{av}$ = 2.108(2) Å.

The zero-field Mößbauer spectra of **6** and **7** are shown in Figs. 2.14 and 2.15.

The spectra of crystalline samples of **6** at 77 K display a quadrupole doublet with an isomer shift $\delta = 0.69(1)$ mm s^{-1}, a quadrupole splitting parameter $\Delta E_Q = 2.27(1)$ mm s^{-1}, and a line width $\Gamma_{FWHM} = 0.48(1)$ mm s^{-1}. These parameters are characteristic of four-coordinate d^6 high-spin ($S = 2$) Fe(II) complexes [17]. Considering the trigonal coordination of the *tris*(carbene) chelator we suggest a ligand field, in which the paired electron is in an e orbital ($e^3 a_1{}^1 e^2$). In contrast, the spectrum of **7** shows a sharp quadrupole doublet ($\Gamma_{FWHM} = 0.26(1)$ mm s^{-1}) with very unusual Mößbauer parameters of

2.2 Results and Discussion

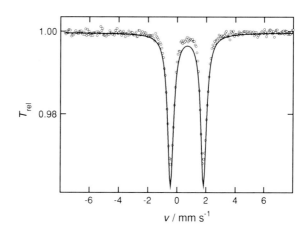

Fig. 2.14 Zero-field Mößbauer spectrum of [(TIMEN^{mes})Fe(N$_3$)]BPh$_4$ (**6**) (recorded in solid state at 77 K; $\delta = 0.69(1)$ mm s^{-1}, $\Delta E_Q = 2.27(1)$ mm s^{-1}, $\Gamma_{FWHM} = 0.48(1)$ mm s^{-1})

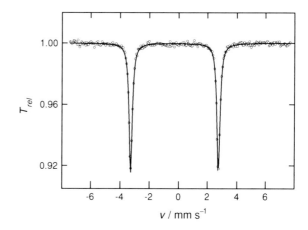

Fig. 2.15 Zero-field Mößbauer spectrum of [(TIMEN^{mes})Fe(N)]BPh$_4$ (**7**) (recorded in solid state at 77 K; $\delta = -0.27(1)$ mm s^{-1}, $\Delta E_Q = 6.04(1)$ mm s^{-1}, $\Gamma_{FWHM} = 0.26(1)$ mm s^{-1})

$\delta = -0.27(1)$ mm s^{-1} and $\Delta E_Q = 6.04(1)$ mm s^{-1}. The negative isomer shift and the remarkably large quadrupole splitting are consistent with values expected for a d^4, $S = 0$ Fe(IV)≡N species. In fact, the observed $\Delta E_Q = 6.04(1)$ mm s^{-1} for **7** is very similar to that determined for the [(PhB^{iPr}P$_3$)Fe(N)] system ($\Delta E_Q = 6.01(1)$ mm s^{-1}) [3]. In both cases, the valence and covalency contributions from the dominating Fe≡N moiety affect the unusually large quadrupole splitting parameter ΔE_Q of the complexes in accordance with the observed and calculated diamagnetic $\{(d_{xy})^2(d_{x^2-y^2})^2\}$ $\{(d_{z^2})0(d_{xz})0(d_{yz})0\}$ ground state electronic configuration that has been computed exemplarily for the 2,6-xylyl derivative (Fig. 2.16).

The positive electron field gradient (EFG) derived from the applied field Mößbauer spectrum (Fig. 2.17) confirms the $\{(d_{xy})^2(d_{x^2-y^2})^2\}$ electron configuration and the asymmetry parameter $\eta = 0$, a measure of deviations from axial symmetry (e.g. towards tetragonal), confirms the three fold symmetry of the

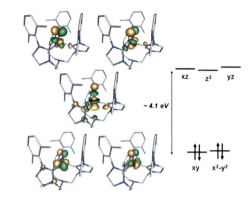

Fig. 2.16 Electronic structure of [(TIMEN^{2,6xyl})Fe(N)]BPh$_4$, $S = 0$ (DFT, ORCA, B3LYP). Computed structural parameters (BP86): Fe–N1 3.278 Å, Fe–N8 1.517 Å, Fe–C$_{av}$ 1.949 Å; ∠(N8–Fe–C) 103.18°

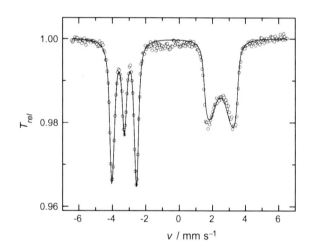

Fig. 2.17 Applied Field Mößbauer spectrum of [(TIMEN^{mes})Fe(N)]BPh$_4$ (**7**) (recorded in solid state at 20 K, 6.5 T; $\eta = 0$)

molecule. In contrast to the striking similarity of the ΔE_Q values, the isomer shift $\delta = -0.27(1)$ mm s^{-1} of **7** differs significantly from that reported for [(PhB^{iPr}P$_3$)Fe(N)] ($\delta = -0.34$ mm s^{-1}). Table 2.1 gives an overview of Mößbauer parameters obtained from the TIMEN^{mes} iron complex series.

The isomer shift determined by ^{57}Fe Mößbauer spectroscopy is directly related to the s electron density at the nucleus, which is shielded by electron density in the shells with nonzero angular momenta (p and d orbitals). Accordingly, the less negative δ indicates that the iron ion in **7** is less oxidized than the tetrahedrally coordinated iron center in [(PhB^{iPr}P$_3$)Fe(N)]. This is likely a consequence of a certain degree of ligand-to-metal back bonding from the NHC ligand's π-system to empty iron d orbitals in trigonal-pyramidal **7**. While it is intuitive that the electronic structure of the four-coordinate iron nitride complex is dominated by the strong Fe≡N interaction, the different isomer shift clearly demonstrates that other factors, such as ligand-donor set and coordination geometry, ultimately determine the detailed electronic structure and the reactivity of high-valent iron nitride

2.2 Results and Discussion

Table 2.1 Mößbauer parameters of [(TIMEN^{mes})Fe(Cl)]Cl (**4**), [(TIMEN^{mes})Fe]BPh$_4$ (**5**), [(TIMEN^{mes})Fe(N$_3$)]BPh$_4$ (**6**), [(TIMEN^{mes})Fe(N)]BPh$_4$ (**7**)

	δ (mm s^{-1})	ΔE_Q (mm s^{-1})	Γ_{FWHM} (mm s^{-1})
[(TIMEN^{mes})Fe(Cl)]Cl (**4**)	0.71(1)	1.78(1)	0.32(1)
[(TIMEN^{mes})Fe]BPh$_4$ (**5**)	0.64(1)	2.34(1)	0.29(1)
[(TIMEN^{mes})Fe(N$_3$)]BPh$_4$ (**6**)	0.69(1)	2.27(1)	0.48(1)
[(TIMEN^{mes})Fe(N)]BPh$_4$ (**7**)	−0.27(1)	6.04(1)	0.26(1)

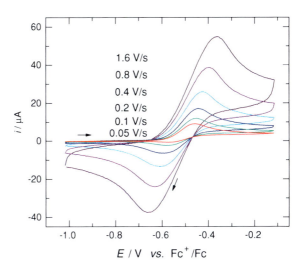

Fig. 2.18 Cyclic voltammogram of [(TIMEN^{mes})Fe(N)]BPh$_4$ (**7**) recorded in THF solution containing 0.1 M TBAPF$_6$ electrolyte under N$_2$; working: glassy carbon, counter: Pt, pseudo-reference: Pt; internal standard: Fc$^+$/Fc; E(Fc$^+$/Fc) = 0.41 V versus NHE

species. Obviously, the neutral, potentially tetradentate *tris*(carbene) ligand and the anionic tridentate *tris*(phosphino)borate, exhibit not only steric but also significant electronic differences. This could also explain the differences in color and stability of the deep purple amino-carbene Fe(IV) nitride [(TIMEN^{mes})Fe(N)]$^+$ (**7**) and the tan-colored, more tetrahedral, [(PhB^{iPr}P$_3$)Fe≡N] complex prone to dimerization. The high stability, not least due to steric shielding of the iron center in **7**, illustrated by the space filling model of the complex (Fig. 2.13), prevents reactivity studies of this molecule. The [FeN] moiety is well protected in the deep cylindrical cavity, thereby completely inhibiting a side access towards the axial ligand. By this means an atom transfer to a substrate is almost entirely prohibited or is only conceivable *via* an axial approach. On the basis of these considerations there are two starting points of increasing the reactivity. On one hand the steric demand of the ligand can be reduced; on the other hand the reactivity can be enhanced *via* manipulating its electronic properties. The redox behaviour of **7** was examined by electrochemical methods.

The cyclic voltammogram of a solution of **7** in THF exhibits a reversible one-electron oxidation at a potential of −0.515 V versus Fc/Fc$^+$ (Fig. 2.18). Fe(IV) nitride complex **7**, dissolved in acetonitrile, was oxidized using different oxidizing agents such as [Fe(Cp)$_2$]PF$_6$, AgOTf and NOBF$_4$ at temperatures below −40 °C.

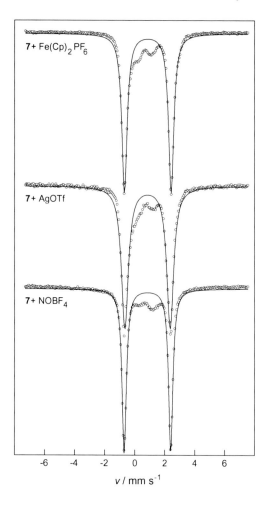

Fig. 2.19 Zero-field Mößbauer spectra of oxidation reaction products of [(TIMEN^{mes})Fe(N)]BPh$_4$ (7) with [Fe(Cp)$_2$]PF$_6$, AgOTf and NOBF$_4$ (^{57}Fe-enriched samples in acetonitrile solution, 77 K; $\delta = 0.85(1)$ mm s^{-1}, $\Delta E_Q = 3.10(1)$ mm s^{-1})

All oxidizers were able to induce an immediate reaction indicated by a spontaneous color change from purple to light green. The reaction mixtures were cooled down to liquid nitrogen temperature after completion of the reaction. Figure 2.19 shows the zero-field Mößbauer spectra of the resulting compounds.

The spectra of the ^{57}Fe enriched samples in acetonitrile at 77 K display a quadrupole doublet with an isomer shift $\delta = 0.85(1)$ mm s^{-1} and a quadrupole splitting parameter $\Delta E_Q = 3.10(1)$ mm s^{-1}. These parameters would match best to a four-coordinate d^6 high-spin ($S = 2$) Fe(II) complex. The solutions of the obtained compounds are temperature sensitive and change color from light green to deep brown when stored at RT. Storage of the solutions at -35 °C slows down the color change but can not prevent the process. According to NMR measurements of the solutions of the resulting products these compounds seem to be NMR silent. After several crystallization attempts, single crystals suitable for an X-ray

2.2 Results and Discussion

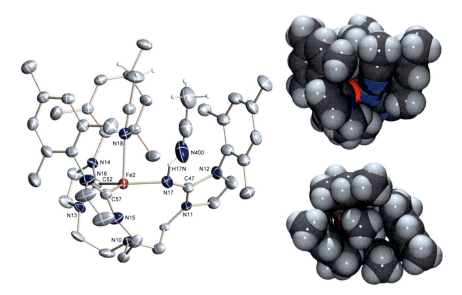

Fig. 2.20 Molecular structure of the complex cation of [(TIMEN^{mes}=N^H)Fe(MeCN)]BPh$_4$ (**8**) in crystals of [(TIMEN^{mes}=N^H)Fe(MeCN)](BPh$_4$)$_2$ · [(TIMEN^{mes}=N^H)Fe(MeCN)](BF$_4$)(BPh$_4$) · Et$_2$O · 1.5 MeCN (30% probability ellipsoids). Hydrogen atoms of the TIMEN^{mes} ligand (except H17A), BPh$_4^-$ anions, BF$_4^-$ anion, and co-crystallized solvents are omitted for clarity. Space-filling representations of the cation **8** *side view* (*right, top*) and *top view* (*right, bottom*). Selected bond distances (Å): Fe2–N18 2.237(3), Fe2–N17 2.005(3), N17–C47 1.341(9), N17–H17 N 1.00, Fe2–C52 2.091(4), Fe2–C57 2.097(4), Fe2–N10 2.585(4), $d_{Fe\ oop}$ 0.110(2)

structure analysis could be obtained at RT by slow diffusion of Et$_2$O into a MeCN solution.

The molecular structure of [(TIMEN^{mes}=N^H)Fe(MeCN)]BPh$_4$ (**8**, Fig. 2.20) shows that the iron center is still coordinated in a trigonal-pyramidal fashion by the ligand, but a nitride insertion has occurred. The axially coordinated nitride ligand of **7** is inserted into one of the Fe–carbene bonds and the axial position is now occupied by an acetonitrile solvent molecule. After this insertion the original *tris*(carbene) ligand possesses two unaltered carbene ligand arms and one imidazol-2-imine moiety. It can be postulated that the reaction mechanism starts with the oxidation of the Fe(IV) nitride to an Fe(V) nitride species, followed by nitride insertion into one of the Fe–carbene bonds (Scheme 2.3). The immediate reaction product is most likely an iron(III) complex. After abstraction of a hydrogen atom, most probably from decomposition reactions of either the solvent or the iron complex, the compound is isolated as an Fe(II) imine species.

The insertion product could not be isolated in reasonable yields and appears to be formed only in trace amounts. The last step of the proposed mechanism, i.e. the hydrogen abstraction, could be the crucial point. If the hydrogen abstraction is hindered or too slow, the newly formed complex cannot be stabilized and undergoes decomposition. Attempts to facilitate the hydrogen abstraction by

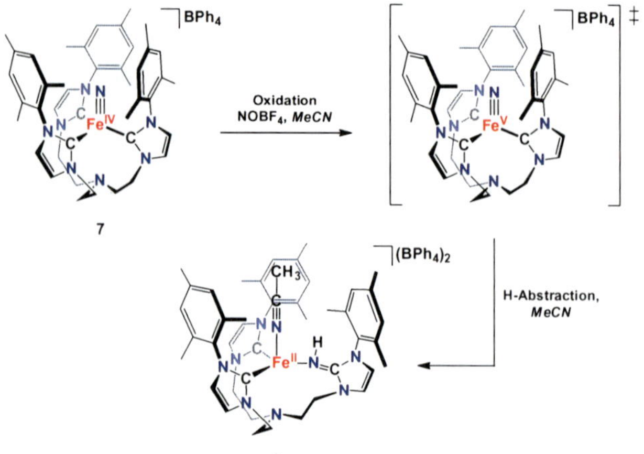

Scheme 2.3 Insertion reaction of [(TIMEN^{mes})Fe(N)]BPh$_4$ (**7**) under oxidative conditions

addition of a hydrogen source, e.g. 1,4-cyclohexadiene, were unsuccessful and did not improve the yield. Therefore, **8** could only be detected by ESI-MS and characterized by X-ray crystal structure determination. The species observed by mass spectrometry is the insertion product, without the axially coordinated ace-tonitrile. There are other examples for insertion reactions with transition metal complexes bearing the TIMEN^R system. Treating [(TIMEN^{mes})Fe]BPh$_4$ (**5**) with an excess of phenyl azide in THF at RT, also yields an insertion product [(TIMEN^{mes}=N^{ph})Fe]BPh$_4$ (**9**, Scheme 2.4). During the reaction, gas evolution could be monitored, while the red solution, stemming from the Fe(I) species, turned brown. Again, the insertion product **9** was formed in low yield and could solely be characterized by X-ray crystal structure determination. The experiment was redone at -35 °C, in order to obtain either the intermediate imido species or to improve the synthesis of the insertion product. Neither the isolation of the intermediate nor the improvement of the insertion reaction was possible by run-ning the experiment at low temperature. Several attempts gave only a few crystals of the compound, thereby precluding an in-depth investigation of the complex. The molecular structure of **9** shows large similarities with **8** (Fig. 2.21; Table 2.2).

As observed for **8**, the TIMEN^{mes}=N^{ph} ligand coordinates to the iron center through two of its remaining carbenoid carbons, the anchoring nitrogen and the newly formed imine nitrogen atom. In contrast to **8**, no axial solvent molecule is bound due to steric hindrance exerted by the phenyl substituent at the imine nitrogen. As a result, when comparing complex **8** and **9** the most significant differences in the molecular structure are visible in the coordination sphere of the iron center. Without axially coordinated acetonitrile, the Fe1–N1 bond in complex **9** is 0.312 Å shorter than in complex **8**. The increase in bond strength is also reflected in the different out-of-plane shifts of **8** and **9**. Whereas in **8** the out-of-plane shift, $d_{Fe\ oop}$ is only 0.097(2) Å and directed towards the axially coordinated acetonitrile, the out-of-plane shift in **9** is much more pronounced and oriented in

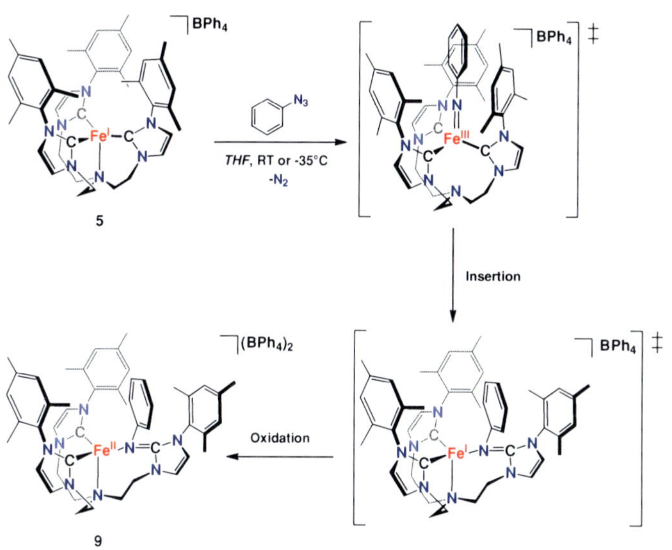

2.2 Results and Discussion

Scheme 2.4 Spontaneous imido insertion into the iron-carbene bond: [(TIMEN^{mes}=N^{ph})Fe](BPh$_4$)$_2$ (**9**)

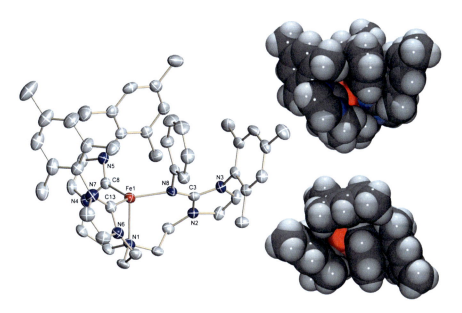

Fig. 2.21 Molecular structure of the complex cation of [(TIMEN^{mes}=N^{ph})Fe](BPh$_4$)$_2$ (**9**) in crystals of **9** · solv (50% probability ellipsoids). Hydrogen atoms, BPh$_4^-$ anions, and co-crystallized solvents are omitted for clarity. Space-filling representations of the cation **9** *side view* (*right, top*) and *top view* (*right, bottom*). Selected bond distances (Å): Fe1–N8 2.017(3), N8–C3 1.338(4), N8–C43 1.456(4), Fe1–C8 2.055(3), Fe1–C13 2.092(3), Fe1–N1 2.244(3), $d_{Fe\ oop}$ 0.222(2)

Table 2.2 Selected bond distances (Å) and bond angles (°) with e.s.d.'s in parentheses for [(TIMEN^{mes}=N^H)Fe(MeCN)](BPh$_4$)$_2$ · [(TIMEN^{mes}=N^H)Fe(MeCN)](BF$_4$)(BPh$_4$) · Et$_2$O · 1.5 MeCN (**8**· Et$_2$O · 1.5 MeCN), [(TIMEN^{mes}=N^{ph})Fe](BPh$_4$)$_2$ · solv (**9** · solv)

	8· Et$_2$O · 1.5 MeCN	**9** · Solv
Fe1–N1	2.556(3)	2.244(3)
Fe1–N8	2.030(3)	2.017(3)
Fe1–N9	2.237(3)	–
N8–C3	1.317(4)	1.338(4)
N8–H8 N/C43	1.00	1.456(4)
Fe1–C8	2.098(3)	2.055(3)
Fe1–C13	2.106(4)	2.092(3)
N2–C3	1.358(4)	1.366(4)
N3–C3	1.363(5)	1.356(4)
N4–C8	1.354(4)	1.350(4)
N5–C8	1.362(4)	1.364(4)
N6–C13	1.360(4)	1.364(4)
N7–C13	1.353(4)	1.348(4)
N8–Fe1–C8	128.02(12)	123.80(12)
N8–Fe1–C13	111.35(13)	113.77(12)
C8–Fe1–C13	119.98(13)	118.93(13)
C3–N8–Fe1	138.7(3)	136.0(2)
H8 N/C43–N8–Fe1	108.6	104.29(19)
C3–N8–H8 N/C43	108.1	117.7(3)
$d_{Fe\ oop}$	0.097(2)	0.222(2)

the opposite direction. Removal of the axial ligand reverses the out-of-plane shift due to a strengthening of the bonding interaction between iron and the nitrogen anchor. In **9**, the remaining iron-carbene bonds, as well as the iron-imine bond are slightly shorter than in **8**. In general, the iron-ligand distances in the four-coordinate complex **9** are shorter than in the five-coordinated complex **8**.

The same type of reaction has been observed before when treating the highly reactive Co(I) complex [(TIMEN^R)Co]Cl with different aryl azides [27]. In this case Co(III) imido complexes [(TIMEN^R)Co(NAr$^{R'}$)]Cl can be isolated at low temperature in near quantitative yields. These complexes are stable in the solid state and in solution at −35 °C. In solution at RT, however, the imido group readily inserts into one of the Co-carbene bonds. Also in these cases, the yields are low and the formation of the insertion products is accompanied by decomposition of a certain amount of the Co(III) imido species.

2.3 Conclusion

In summary, the generation and stabilization of a discrete Fe(IV) nitride complex was accomplished employing a sterically demanding nitrogen-anchored N-heterocyclic *tris*(carbene) ligand (TIMEN^{mes}). The tripodal TIMEN^{mes} ligand was coordinated to iron by treatment with ferrous chloride. The corresponding Fe(II) azide complex **6** could be obtained *via* two different reaction routes: either by reduction of the Fe(II) chloro precursor **4** to the Fe(I) species **5**, followed by treatment of the resulting complex with TMS-azide. Alternatively, by salt metathesis reaction, stirring the Fe(II) precursor **4** with an excess of sodium azide in acetonitrile. Photolysis of the Fe(II) azide complex **6** generates the Fe(IV) nitride complex **7**. The crystal structure determination of this iron nitride complex **7** reported here is the first structural verification of the [FeN] synthon that has been recognized as the reactive intermediate in the industrial Haber–Bosch process. Complex **7** is remarkably air and moisture stable at room temperature, which can be attributed to the steric protection provided by the $TIMEN^{mes}$ ligand. The X-ray crystallographically determined $Fe \equiv N$ bond is short at 1.53 Å, and the high-valent Fe(IV) center in this molecule possesses a trigonal-pyramidal coordination environment. Aiming for high-valent iron N-heterocyclic carbene complexes, and especially for iron nitride complexes, we tried to obtain an iron(V) nitride complex, as well. Attempts to oxidize **7** with different oxidizing reagents yielded the insertion product $[(TIMEN^{mes}=N^H)Fe(MeCN)]BPh_4$ (**8**). Similar reaction behavior can be observed with transition metal imido complexes $[(TIMEN^R)M=N^R]^+$ (M=Co(III), Fe(III)).

2.4 Experimental

2.4.1 Methods, Procedures, and Starting Materials

All air- and moisture-sensitive experiments were performed under dry nitrogen atmosphere using standard Schlenk techniques or an MBraun inert-gas glovebox containing an atmosphere of purified dinitrogen. Iron(II) chloride, anhydrous 99.9%, was purchased from Aldrich and used as received. Trimethylsilyl azide 97% and sodium tetraphenylborate were purchased from ACROS and were used without further purification. Phenyl azide was prepared following the literature procedure [28]. Solvents were purified using a two-column solid-state purification system (Glasscontour System, Irvine, CA) and transferred to the glovebox without exposure to air. NMR solvents were obtained packaged under argon and stored over activated molecular sieves and sodium (where appropriate) prior to use.

^{1}H-NMR spectra were recorded on JEOL 270 and 400 MHz instruments, operating at respective frequencies of 269.714 and 400.178 MHz with a probe temperature of 23 °C.

^{13}C-NMR spectra were recorded on JEOL 270 and 400 MHz instruments, operating at respective frequencies of 67.82 and 100.624 MHz with a probe temperature of 23 °C. Chemical shifts were reported relative to the peak for $SiMe_4$, using ^{1}H (residual) chemical shifts of the solvent as a secondary standard and are reported in ppm.

^{15}N-NMR data were acquired on a JEOL 400 MHz instrument, and chemical shifts were referenced to CH_3NO_2 (380.23 ppm relative to liquid ammonia at 0 ppm).

Electronic absorption spectra were recorded from 300 to 2,000 nm with a UV-vis-NIR-Spectrophotometer (Shimadzu UV-3600) in THF.

All infrared spectra were recorded on a Shimadzu IRPrestige-21 system in KBr pellets.

Elemental analysis results were obtained from the Analytical Laboratories at the Friedrich-Alexander-University Erlangen-Nuremberg (Erlangen, Germany).

^{57}Fe Mößbauer spectra were recorded on a WissEl Mößbauer spectrometer (MRG-500) at 77 K in constant acceleration mode. ^{57}Co/Rh was used as the radiation source. WinNormos for Igor Pro software was used for the quantitative evaluation of the spectral parameters (least-squares fitting to Lorentzian peaks). The minimum experimental line widths were 0.23 mm s^{-1}.

The temperature of the samples was controlled by an MBBC-HE0106 MÖSSBAUER He/N$_2$ cryostat within an accuracy of ±0.3 K. Isomer shifts were determined relative to α-iron at 298 K.

Magnetic data were collected using a Quantum Design MPMS-XL SQUID magnetometer. Measurements were obtained for a finely ground microcrystalline powder restrained within a polycarbonate gel capsule. Dc susceptibility data were collected in the temperature range 2–300 K under a dc field of 1 T. The data were corrected for core diamagnetism of the sample estimated using Pascal's constants. The program julX written by E. Bill was used for the simulation and analysis of magnetic susceptibiltiy data [18].

EPR measurements were performed in quartz tubes with J. Young valves. Frozen solution EPR spectra were recorded on a JEOL continuous wave spectrometer JESFA200 equipped with an X-band Gunn diode oscillator bridge, a cylindrical mode cavity, and a helium cryostat. All spectra were obtained on freshly prepared solutions of 1–10 mM compound in acetonitrile. The spectra were simulated using the W95EPR program [29].

Samples for mass spectrometry were prepared by dissolving the compound in dry acetonitrile under nitrogen to yield a 10^{-4} M solution. Using a syringe pump at a flow rate of 240 mL h^{-1}, the acetonitrile solutions were infused into an orthogonal ESI source of an Esquire 6,000 ion trap mass spectrometer (Bruker, Bremen, Germany). Nitrogen was used as the nebulizing gas. The source voltages for the positive-ion mode were set as follows: capillary at -4.0 kV, end plate offset at -500 V, capillary exit at 151.0 V. The ion trap was optimized for the respective target mass.

2.4.2 Computational Details

The electronic structure and geometry of the iron(IV) nitride complex **7** were investigated with Kohn–Sham DFT calculations using the ORCA package [30]. The B3LYP [31–33] functional was used together with the SV(P) basis set [34] for the main group elements, and the TZV(P) basis for the iron centre [35].

2.4 Experimental   43

2.4.3 Synthetic Details

Im^{mes} (1). A solution of ammonium acetate (30.8 g, 0.4 mol) and 2,4,6-trimeth-ylaniline (54.1 g, 51.9 mL, 0.4 mol) dissolved in 100 mL acetic acid was added dropwise to a stirred solution of glyoxal (58.1 g, 45.9 mL, 0.4 mol) and formaldehyde (32.7 g, 30.0 mL, 0.4 mol) in 100 mL acetic acid at 80 °C. The resulting mixture was stirred for 16 h at 80 °C during which the solution turned brown. The reaction mixture was added dropwise to a stirred solution of KOH in distilled water (200 g KOH in 1.25 L), the crude product was extracted with methylene chloride. After drying with sodium sulfate, filtration, and removing of solvent the residue was purified by vacuum distillation (36.8 g; yield 49.4%). ^{1}H-NMR (270 MHz, chloroform-d_1, RT): δ (ppm) = 7.78 (s, 1H), 7.35 (s, 1H), 7.20 (s, 1H), 6.94 (s, 2H), 2.30 (s, 3H), 1.94 (s, 6H). ^{13}C{^{1}H}-NMR (68 MHz, chloroform-d_1): δ (ppm) = 137.1, 135.3, 129.0, 128.7, 128.4, 120.0, 20.9, 17.3.

[H$_3$TIMEN^{mes}](PF$_6$)$_3$ (2). A suspension of *tris*(2-chlorethyl)amine-hydro-chloride (9.00 g, 37.3 mmol) and potassium hydroxid (3.28 g, 56.7 mmol, ca. 1.5 eq.) was stirred for 30 min until all hydrochloride salt had separated from the aqueous solution. The light yellow oil was extracted with methylene chloride (3 × 20 mL) and the solvent was removed using the rotovap (7.30 g; yield 95.7%). A 100 mL flask was charged with *tris*(2-chloroethyl)amine (7.30 g, 35.7 mmol) and **1** (19.95 g, 0.11 mol), and the mixture was heated to 150 °C for 2 days. During this time, the reaction mixture became a brown solid, to complete the conversion the solidified melt can be redissolved by adding acetonitrile from time to time. After 2 days the resulting solid was dissolved in a minimum amount of hot methanol. The hygroscopic chloride salt [H$_3$TIMEN^{mes}](Cl)$_3$ (**2a**) in solution, was converted to the corresponding stable hexafluorophosphate salt, [H$_3$TIMEN^{mes}](PF$_6$)$_3$, by addition of a solution of NH$_4$PF$_6$ (20.00 g, 12.27 mol) in a minimum amount of hot methanol. The off-white hexafluorophosphate salt was collected by filtration, washed with methanol and diethyl ether. The resulting solid was dried in vacuo (23.39 g; yield 60%). ^{1}H-NMR (270 MHz, DMSO-d_6, RT): δ (ppm) = 9.37 (s, 3H), 8.01 (d, 6H), 7.16 (s, 6H), 4.36 (t, ^{3}J(H,H) 4.45 Hz, 6H), 3.13 (t, ^{3}J(H,H) 4.45 Hz, 6H), 2.32 (s, 9 H), 2.00 (s, 18H). ^{13}C {^{1}H}-NMR (68 MHz, DMSO-d_6, RT): δ (ppm) = 215.0, 140.4, 137.4, 134.3, 131.1, 124.0, 123.4, 52.1, 46.7, 20.6, 16.9.

[TIMEN^{mes}] (3). A suspension of potassium *tert*-butoxide (0.45 g, 4.04 mmol) and [H$_3$TIMEN^{mes}](PF$_6$)$_3$ (**2**) (1.43 g, 1.30 mmol) in 20 mL Et$_2$O was stirred for 5 h. The solution was filtered over celite and the filtrate was evaporated to dryness in vacuo. ^{1}H-NMR (400 MHz, benzene-d_6, RT): δ (ppm) = 6.80 (d, ^{3}J(H,H) 1.62 Hz, 3H), 6.71 (s, 6H), 6.28 (d, ^{3}J(H,H) 1.62 Hz, 3H), 4.02 (t, ^{3}J(H,H) 6.53 Hz, 6H), 2.85 (t, ^{3}J(H,H) 6.53 Hz, 6H), 2.06 (s, 9H), 2.04 (s, 18H). ^{13}C {^{1}H}-NMR (100 MHz, benzene-d_6, RT): δ (ppm) = 213.59, 150.29, 139.11, 137.19, 135.38, 134.88, 123.17, 120.19, 120.07, 56.62, 49.68, 20.95, 18.13.

[(TIMEN^{mes})Fe(Cl)]Cl (4). A solution of FeCl$_2$ (0.17 g, 1.3 mmol) in 10 mL pyridine was added to a solution of TIMEN^{mes} (**3**) (0.85 g, 1.30 mmol) in 10 mL

Fig. 2.22 ^{1}H-NMR spectrum of [(TIMEN^{mes})Fe(N$_3$)]BPh$_4$ **6** (recorded in DMSO-d_6 at RT)

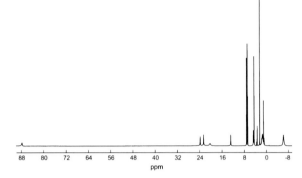

pyridine. The reaction mixture was allowed to stir overnight, during which an off-white precipitate formed. The precipitate was collected by filtration, washed with pyridine, diethyl ether and n-pentane and dried in vacuo (1.02 g; yield 80%). ^{1}H-NMR (270 MHz, CD$_2$Cl$_2$, RT): δ (ppm) = 77.14 (s, 3H), 25.45 (s, 3H), 20.12 (s, 3H), 19.53 (s, 3H), 9.99 (s, 3H), 8.55 (s, 3H), 5.14 (s 9H), 4.96 (s, 3H), 1.79 (s, 3H), −0.32 (s, 9H), −5.90 (s, 9H). Elemental analysis (%) for C$_{42}$H$_{51}$Cl$_2$FeN$_7$ calcd. C 64.62, H 6.58, N 12.56; obsd. C 64.78, H 6.50, N 12.84. Mößbauer (solid state, 80 K) δ = 0.71(1) mm s^{-1}, ΔE_Q = 1.78(1) mm s^{-1}, Γ_{FWHM} = 0.32(1) mm s^{-1}. SQUID (χ_{dia} = 4.53 × 10^{-4} cm^3 mol^{-1}, RT) μ_{eff} = 4.89 μ_B.

[(TIMEN^{mes})Fe]BPh$_4$ (5). A mixture of [(TIMEN^{mes})Fe(Cl)]Cl (**4**) (0.50 g, 0.64 mmol) and sodium tetraphenylborate (0.22 g, 0.64 mmol) in 20 mL THF was stirred over sodium amalgam (5.00 g, equates to 22.99 mg, 1.70 mmol sodium) overnight. The solution was filtered over celite, concentrated and cooled to −35 °C to yield the product as dark red crystals (0.53 g, yield 80%). Mößbauer parameters (solid state, 80 K) δ = 0.64(1) mm s^{-1}, ΔE_Q = 2.34(1) mm s^{-1}, Γ_{FWHM} = 0.29(1) mm s^{-1}. EPR (acetonitrile, 14 K, 8.962 GHz, Mod-Width = 1.0 mT, Power = 0.998 mW, $g_\perp$ = 4.2, $g_\parallel$ = 2.025, $\Delta H_{1/2\perp}$ = 6.7 mT, $\Delta H_{1/2\parallel}$ = 4.0 mT.

[(TIMEN^{mes})Fe(N$_3$)]BPh$_4$ (6). A: Excess TMS-N$_3$ was added to a solution of [(TIMEN^{mes})Fe]BPh$_4$ (**5**) (195.50 mg, 0.19 mmol) in 20 mL THF. The reaction mixture was stirred overnight, during which the red solution turned yellow. Half of the solvent was removed and addition of 10 mL of diethyl ether to the solution induced precipitation of a light yellow powder. The precipitate was filtered, washed with diethyl ether and pentane, and dried in a vacuum (161.0 mg; yield: 79%). Light yellow crystals suitable for X-ray diffraction analysis were grown by diffusion of pentane into a saturated solution of **6** in THF at room temperature. **B:** A mixture of [(TIMEN^{mes})Fe(Cl)]Cl (**4**) (0.40 g, 0.51 mmol), sodium tetraphenylborate (0.17 g, 0.51 mmol) and sodium azide (0.07 g, 1.10 mmol) was stirred in 15 mL THF for 2 days. The solution was filtered over celite and the solvent removed in vacuo. In the ^{1}H-NMR, eleven signals are expected for the complex, nine are observed, two signals are likely broadened into the baseline. ^{1}H-NMR (400 MHz, DMSO-d_6, RT): δ (ppm) = 87.50 (s, 3H), 23.80 (s, 3H), 22.58 (s, 3H), 20.28 (s, 3H), 12.89 (s, 3H), 7.18

2.4 Experimental 45

(m, 8H), 6.94 (m, 8H), 6.92 (m, 4H), 4.80 (s, 3H), 4.52 (s, 9H), 1.50 (s, 9H), −6.11 (s, 9H). Elemental Analysis (%) for $C_{66}H_{71}BFeN_{10}$ calcd. C 74.02, H 6.68, N 13.08; obsd. C 74.13, H 6.73, N 12.66. Mößbauer (solid state, 80 K) $\delta = 0.687(1)$ mm s^{-1}, $\Delta E_Q = 2.267(3)$ mm s^{-1}, $\Gamma_{FWHM} = 0.48(1)$ mm s^{-1} (Fig. 2.22).

[(TIMEN^{mes})Fe(N)]BPh$_4$ (7). A stirred solution of **6** (161.0 mg, 0.15 mmol) in 20 mL THF was irradiated for 24 h (Osram XBO 150 W Xenon arc lamp) during which the yellow solution turned purple. The solution was filtered through Celite, concentrated and addition of n-pentane to the solution induced precipitation of a purple powder. The precipitate was filtered, washed with n-pentane, and dried in a vacuum (101.9 mg; yield: 65%). Purple crystals suitable for X-ray diffraction analysis were grown by diffusion of diethyl ether into a saturated solution of **7** in THF at room temperature. ^{1}H-NMR (400 MHz, DMSO-d_6, RT): δ (ppm) = 7.59 (d, ^3J(H,H) = 1.8, 3H), 7.20 (d, ^3J(H,H) = 1.8, 3H), 7.16 (m, 8H), 6.86 (m, 8H), 6.75 (m, 4H), 6.61 (s, 3H), 6.60 (s, 3H), 4.06 (m, 3H), 3.10 (m, 3H), 2.68 (m, 3H), 2.26 (m, 3H), 2.13 (s, 9H), 1.61 (s, 9H), 0.92 (s, 9H). ^{13}C-NMR (100 MHz, DMSO-d_6, RT): δ (ppm) = 198.29, 163.94, 138.22, 137.49, 137.29, 136.08, 133.27, 128.57, 128.43, 125.82, 124.63, 122.06, 58.38, 50.56, 20.88, 17.36, 16.80. Elemental analysis (%) for $C_{66}H_{71}BFeN_8$ calcd. C 76.00, H 6.86, N 10.74; obsd. C 75.47, H 7.04, N 10.61. Mößbauer (solid state, 80 K) $\delta = -0.27(1)$ mm s^{-1}, $\Delta E_Q = 6.04(1)$ mm s^{-1}, $\Gamma_{FWHM} = 0.26(1)$mm s^{-1}

[(TIMEN^{mes}=N^H)Fe(MeCN)]BPh$_4$ (8). A solution of NOBF$_4$ (6.1 mg, 0.53 mmol) in 5 mL acetonitrile was added dropwise to a solution of **7** (55.0 mg, 0.53 mmol) in 5 mL of acetonitrile. After complete addition of the oxidizing agent, the solvent was immediately removed in vacuo. Suitable single crystals for X-ray crystal structure determination could be obtained by dissolving the resulting residue in acetonitrile-d_3, filtration and slow diffusion of Et$_2$O into the saturated solution of **8** at RT. ESI-MS (m/z) for $C_{42}H_{52}FeN_8$ calcd. 362 [M^{2+}]; obsd. 362 [M^{2+}].

[(TIMEN^{mes}=N^{ph})Fe]BPh$_4$ (9). An excess of phenyl azide was added to a stirred solution of **5** (50.0 mg, 48.6 μmol) in 10 mL THF. The reaction mixture was stirred for 2 h during which time the red solution became brown and a precipitate was formed, accompanied by gas evolution. Half of the solvent was removed under vacuum and complete precipitation was induced by addition of n-pentane to the mixture. The precipitate was collected by filtration and washed with n-pentane. Suitable single crystals for X-ray crystal structure determination could be obtained by dissolving the resulting residue in a mixture of THF and acetonitrile, filtration and slow diffusion of n-pentane into the saturated solution of **9** at RT.

2.4.4 X-ray Crystal Structure Determination Details

Orange needles of [H$_3$TIMEN^{mes}](Cl)$_3$ (**2a**) were obtained at RT from a saturated solution of **2a**. Orange blocks of [(TIMEN^{mes})Fe(Cl)]Cl · 3 MeCN (**4 · 3 MeCN**) were grown at RT by layering a MeCN solution of **4** with Et$_2$O, ruby red needles of [(TIMEN^{mes})Fe]BPh$_4$ · 2 THF · 0.5 Et$_2$O (**5 · 2 THF · 0.5 Et$_2$O**) were obtained at

RT by slow diffusion of Et_2O into a THF solution of **5**, pale-yellow needle-shaped crystals of $[(TIMEN^{mes})Fe(N_3)]BPh_4 \cdot 2.5$ THF (**6** · 2.5 THF), were grown by layering a THF solution of **6** with n-pentane, dark-violet prisms of $[(TIMEN^{mes})Fe(N)]BPh_4 \cdot Et_2O$ (**7** ·Et_2O) were obtained by slow diffusion of Et_2O into a THF solution of **7** that was layered with n-pentane. Green-brown plates of $[(TIMEN^{mes}=N^H)Fe(MeCN)](BPh_4)_2 \cdot [(TIMEN^{mes}=N^H)Fe(MeCN)](BF_4)(BPh_4) \cdot Et_2O \cdot 1.5$ MeCN (**8** · Et_2O · 1.5 MeCN) were obtained at RT by slow diffusion of Et_2O into a MeCN solution of **8**. Brown plates of $[(TIMEN^{mes}=N^{ph})Fe](BPh_4)_2 \cdot$ solv (**9** · solv; solv = 1.044 n-pentane · 1.023 THF) were obtained at RT by slow diffusion of n-pentane into a THF/acetonitrile solution of **9**. Suitable single crystals were embedded in protective perfluoropolyalkylether oil and transferred to the cold nitrogen gas stream of the diffractometer. Intensity data were collected either on a Bruker-Nonius KappaCCD diffractometer (**5** · 2 THF · 0.5 Et_2O, **6** · 2.5 THF, **7** · Et_2O, **8** · Et_2O · 1.5 MeCN, **9** · solv and **2a**) using graphite monochromatized MoK_α radiation ($\lambda = 0.71073$ Å) or on a Bruker Kappa APEX2 Duo diffractometer (**4** · 3 MeCN) equipped with an $I\mu S$ microsource and QUAZAR focusing optics using MoK_α radiation ($\lambda = 0.71073$ Å). Data were corrected for Lorentz and polarization effects, semiempirical absorption corrections were performed on the basis of multiple scans using *SADABS* [36]. All structures were solved by direct methods and refined by full-matrix least-squares procedures on F^2 using *SHELXTL NT* 6.12 [37]. All non-hydrogen atoms were refined with anisotropic displacement parameters. The hydrogen atoms were placed in positions of optimized geometry, their isotropic displacement parameters were tied to those of the corresponding carrier atoms by a factor of either 1.2 or 1.5. In the crystal structure of **2a** the molecule lies on a crystallographic threefold axis, accordingly the molecule exhibits C_3 symmetry. Compound **4** crystallizes with a total of three molecules of MeCN per formula unit. The residual electron density maximum at the iron center indicates the presence of a small amount of $[(TIMEN^{mes})Fe(MeCN)]^{2+}$ (less than 5%). Adopting the concept of split positions was not successful in this case. Compound **5** crystallizes with a total of two molecules of THF and 0.5 Et_2O per formula unit. One of the THF molecules is disordered. Two alternative orientations were refined in each case resulting in site occupancies of 57.3(5)% for the atoms O200-C204 and 42.7(5)% for the atoms O210-C214, respectively. The half molecule of Et_2O is disordered and situated on a crystallographic inversion center. SIMU and SAME restraints were applied in the refinement of the disordered structure parts. The crystal structure of $[(TIMEN^{mes})Fe(N)]BPh_4 \cdot Et_2O$ (**7** · Et_2O) is subjected to heavy disorder which affects the cationic iron complex exhibiting two different orientations of the chelating ligand around the central atom, the BPh_4 anion having a threefold disorder of the phenyl ring pointing towards the cation as well as the Et_2O solvent molecule that is also threefold disordered. This disorder gives rise to two apparent short contacts between disordered fragments of different parts and orientations, respectively, namely the ligand of the cationic complex and the Et_2O solvent molecules. ISOR and SIMU restraints were used in the refinement of some of the affected atoms. The orientational disorder of the rotor-like arranged ligand leads to two preferred orientations that are occupied by 67.6(2) and 32.4(2)%, respectively, with a few

Table 2.3 Crystallographic data, data collection and refinement details of [(TIMEN^{mes})Fe(Cl)]Cl · 3 MeCN (**4** · 3 MeCN), [(TIMEN^{mes})Fe]BPh$_4$ · 2 THF · 0.5 Et$_2$O (**5** · 2 THF · 0.5 Et$_2$O), [(TIMEN^{mes})Fe(N$_3$)]BPh$_4$ · 2.5 THF (**6** · 2.5 THF), [(TIMEN^{mes})Fe(N)]BPh$_4$ · Et$_2$O (**7** · Et$_2$O)

	4 · 3 MeCN	**5** · 2 THF · 0.5 Et$_2$O	**6** · 2.5 THF	**7** · Et$_2$O
Empirical formula	C$_{48}$H$_{60}$Cl$_2$FeN$_{10}$	C$_{76}$H$_{92}$BFeN$_7$O$_{2.5}$	C$_{76}$H$_{91}$BFeN$_{10}$O$_{2.5}$	C$_{70}$H$_{81}$BFeN$_8$O
Mol. weight	903.81	1,210.23	1,251.25	1,117.09
Crystal size (mm)	0.28 × 0.10 × 0.08	0.50 × 0.14 × 0.13	0.58 × 0.14 × 0.11	0.33 × 0.27 × 0.14
Temperature (K)	100	150	100	100
Crystal system	Triclinic	Triclinic	Triclinic	Triclinic
Space group	$P\bar{1}$ (no. 2)	$P\bar{1}$ (no. 2)	$P\bar{1}$ (no. 2)	$P\bar{1}$ (no. 2)
a (Å)	10.6405(4)	11.2380(10)	11.3035(5)	12.432(2)
b (Å)	10.6824(4)	17.4660(10)	17.725(2)	12.632(2)
c (Å)	23.9778(11)	17.6820(10)	17.831(2)	22.614(3)
α (°)	93.3580(10)	85.287(5)	88.056(10)	80.104(13)
β (°)	93.8020(10)	79.615(6)	76.821(6)	84.734(9)
γ (°)	118.7900(10)	77.372(5)	78.047(7)	61.303(7)
V (Å^3)	2,370.51(17)	3,327.9(4)	3,402.7(6)	3,068.6(8)
Z	2	2	2	2
ρ (g cm^{-3}) (calc.)	1.266	1.208	1.221	1.209
μ (mm^{-1})	0.475	0.279	0.277	0.296
F (000)	956	1,296	1,336	1,192
T_{min}; T_{max}	0.688; 0.746	0.826; 0.960	0.861; 0.970	0.846; 0.960
2Θ interval (°)	$5.62 \leq 2\Theta \leq 54.2$	$6.24 \leq 2\Theta \leq 54.2$	$6.6 \leq 2\Theta \leq 52.8$	$6.9 \leq 2\Theta \leq 54.2$
Coll. refl.	46,788	94,905	66,209	69,261
Indep. refl.	10,377	15,670	13,885	13,497
Obs. refl. $F_0 \geq 4\sigma(F)$	8,227	10,608	9,649	9,016
No. ref. param.	562	855	847	1,346
wR_2 (all data)	0.1398	0.1496	0.1292	0.0938
R_1 ($F_0 \geq 4\sigma(F)$)	0.0522	0.0560	0.0516	0.0450
GooF on F^2	1.068	1.020	1.021	1.042
Max.; min. residual electron density	1.377; −0.533	1.263; −0.549	0.623; −0.422	0.305; −0.275

48   2 TIMEN^{mes}: An Iron Nitride Complex

Table 2.4 Crystallographic data, data collection and refinement details of [(TIMEN^{mes}=N^H)Fe(MeCN)](BPh$_4$)$_2$ · [(TIMEN^{mes}=N^H)Fe(MeCN)](BF$_4$)(BPh$_4$) · Et$_2$O · 1.5 MeCN (**8**· Et$_2$O · 1.5 MeCN), [(TIMEN^{mes}=N^{ph})Fe](BPh$_4$)$_2$ · solv (**9** · solv), [H$_3$TIMEN^{mes}](Cl)$_3$ (**2a**)

	8 · Et$_2$O · 1.5 MeCN	**9** · solv	**2a**
Empirical formula	C$_{167}$H$_{184.5}$B$_4$F$_4$Fe$_2$N$_{19.5}$O$_{1.02}$	C$_{105.31}$H$_{116.28}$B$_2$FeN$_8$O$_{1.02}$	C$_{42}$H$_{54}$Cl$_3$N$_7$
Mol. weight	2,711.78	1,587.92	1,251.25
Crystal size	0.36 × 0.24 × 0.10	0.40 × 0.18 × 0.05	0.62 × 0.18 × 0.18
Temperature (K)	200	100	150
Crystal system	Monoclinic	Monoclinic	Trigonal
Space group	$P2_1/c$ (no. 14)	$P2_1/c$ (no. 14)	$R\bar{3}$ (no. 148)
a (Å)	15.570(2)	29.432(2)	21.687(3)
b (Å)	18.196(1)	14.6980(15)	21.687(3)
c (Å)	47.439(9)	21.1561(15)	15.1534(9)
α (°)	90	90	90
β (°)	98.598(10)	107.317(6)	90
γ (°)	90	90	120
V (Å^3)	14,996(3)	8,737.1(12)	6,172.2(13)
Z	4	4	6
ρ (g cm^{-3}) (calc.)	1.201	1.207	1.232
μ (mm^{-1})	0.258	0.228	0.261
F (000)	5,756	3,393	2,436
T_{min}; T_{max}	0.874; 0.980	0.871; 0.990	0.797; 0.950
2Θ interval (°)	$6.46 \leq 2\Theta \leq 50.06$	$6.42 \leq 2\Theta \leq 51.36$	$6.5 \leq 2\Theta \leq 54.2$
Coll. refl.	169,822	119,834	44,209
Indep. refl.	26,372	16,559	3,026
Obs. refl. $F_0 \geq 4\sigma(F)$	15,098	10,886	2,374
No. ref. param.	2,025	1,175	160
wR_2 (all data)	0.1680	0.1871	0.1045
R_1 ($F_0 \geq 4\sigma(F)$)	0.0682	0.0684	0.0379
GooF on F^2	1.066	1.040	1.093
Max.; min. residual electron density	0.610; −0.628	1.232; −0.590	0.389; −0.252

atoms belonging to both orientations; most importantly, the amine nitrogen, N1, the iron ion, Fe1, and the nitride, N8, as well as N2, N4, and C1. For the disordered phenyl ring (C61–C66) of the BPh$_4$ anion three preferred orientations could be refined that are occupied by 36.0(5)% for the atoms C62–C66, 30.5(5)% for the atoms C62A–C66A, and 34.0(5)% for the atoms C62B–C66B. The treatment of the threefold disorder of the Et$_2$O solvent molecule resulted in site occupancies of 38.0(4)% for C101–C105, 32.0(4)% for C201–C205, and 30.0(4)% for C301–C305. A number of restraints (SIMU, ISOR, and SAME) was applied in the refinement of the disordered regions of the crystal structure of **7** · Et$_2$O. The crystal structure of [(TIMEN^{mes}=N^H)Fe(MeCN)](BPh$_4$)$_2$ · [(TIMEN^{mes}=N^H)Fe(MeCN)](BF$_4$)(BPh$_4$) · Et$_2$O 1.5 MeCN (**8** · Et$_2$O · 1.5 MeCN) contains two independent molecules of the complex cation in the asymmetric unit, one of which (Fe2) is disordered. Two

2.4 Experimental

Table 2.5 Selected bond distances (Å) and bond angles (°) with e.s.d.'s in parentheses for [(TIMEN^{mes})Fe(Cl)]Cl · 3 MeCN (**4** · 3 MeCN), [(TIMEN^{mes})Fe]BPh$_4$ · 2 THF · 0.5 Et$_2$O (**5** · 2 THF · 0.5 Et$_2$O), [(TIMEN^{mes})Fe(N$_3$)]BPh$_4$ · 2.5 THF (**6** · 2.5 THF), [(TIMEN^{mes})Fe(N)]BPh$_4$ · Et$_2$O (**7** · Et$_2$O)

	4 · 3 MeCN	**5** · 2 THF · 0.5 Et$_2$O	**6** · 2.5 THF	**7** · Et$_2$O (major)[a]	**7** · Et$_2$O (minor)[a]
Fe1–N1	3.008(3)	2.275(2)	3.244(2)		3.120(2)
Fe1–Lig$_{axial}$[b]	2.2768(8)	–	1.947(2)		1.526(2)
Fe1–C3	2.128(3)	2.032(2)	2.104(2)	1.987(3)	1.920(7)
Fe1–C8	2.138(3)	2.024(2)	2.110(3)	1.955(5)	1.982(10)
Fe1–C13	2.138(2)	2.012(2)	2.111(2)	1.946(4)	2.006(8)
N8–N9	–	–	1.183(3)	–	–
N9–N10	–	–	1.157(3)	–	–
N2–C3	1.360(3)	1.378(3)	1.363(3)	1.323(4)	1.391(7)
N3–C3	1.376(3)	1.380(3)	1.365(3)	1.382(4)	1.370(8)
N4–C8	1.365(3)	1.370(3)	1.367(3)	1.361(5)	1.268(10)
N5–C8	1.360(3)	1.374(3)	1.362(3)	1.373(5)	1.359(10)
N6–C13	1.359(3)	1.374(3)	1.356(3)	1.409(9)	1.31(2)
N7–C13	1.371(3)	1.374(3)	1.368(3)	1.374(4)	1.375(9)
Fe1–N8–N9	166.9(2)	–	174.5(2)	–	–
N8–N9–N10	179.1(2)	–	179.6(3)	–	–
C3–Fe1–Lig$_{axial}$	100.56(7)	–	103.9(1)	115.1(2)	101.5(2)
C8–Fe1–Lig$_{axial}$[b]	102.41(7)	–	105.2(1)	101.4(2)	103.4(3)
C13–Fe1–Lig$_{axial}$[b]	102.61(7)	–	106.6(1)	103.8(2)	102.0(2)
$d_{Fe\ oop}$	0.439(2)	0.191(2)	0.554(2)	0.427(3)	0.420(5)

[a] In its crystal structure the cation **7** exhibits two different orientations of the chelate ligand around the central atom with the major component being occupied by 67.6(2) and the minor component by 32.4(2)%

[b] Lig$_{axial}$ corresponds to the respective axial Ligand, for compound **4**: Cl (chloro); for compound **6**: N8 (azide); for compound **7**: N8 (nitride)

alternative positions were refined in the case of the ligand arm which is affected by insertion, each position is refined with occupancies of 50%. The BF$_4^-$ anion is disordered over two sites with refined occupancies of 62.0(3)% and 38.0(3)% for the atoms B1–F14 and B1A–F14A. The solvent molecule Et$_2$O is also disordered, whereas the 50:50-occupancy is correlated with the disorder of the cation (Fe2). Also the occupancy of 50% of one of the MeCN molecules is based on the disorder of the cation (Fe2). The crystal structure of [(TIMEN^{mes}=N^{ph})Fe](BPh$_4$)$_2$ · solv (**9** · solv; solv = 1.044 n-pentane · 1.023 THF) contains several highly disordered solvent molecules (THF and n-pentane). In parts different solvent molecules share one crystallographic site, which results in a non-integer sum formula. SIMU, ISOR, DFIX and SAME restraints were applied in the refinement of the disordered solvent molecules (Tables 2.3, 2.4, 2.5).

50 2 TIMEN^{mes}: An Iron Nitride Complex

Acknowledgement Text, schemes, and figures of this chapter, in part, are reprints of the materials published in Vogel et al. [38]. The dissertation author was the primary researcher and author. The co-authors listed in the publication also participated in the research. The permission to reproduce the paper was granted by Wiley-VCH Verlag GmbH & Co. KGaA. Copyright 2008, Wiley-VCH Verlag GmbH & Co. KGaA, Weinheim.

References

1. J.T. Groves, J. Inorg. Biochem. **100**, 434 (2006)
2. O. Einsle, F.A. Tezcan, S.L.A. Andrade, B. Schmid, M. Yoshida, J.B. Howard, D.C. Rees, Science **297**, 1696 (2002)
3. M.P. Hendrich, W. Gunderson, R.K. Behan, M.T. Green, M.P. Mehn, T.A. Betley, C.C. Lu, J.C. Peters, Proc. Natl. Acad. Sci. **103**, 17107 (2006)
4. J.F. Berry, E. Bill, E. Bothe, S.D. George, B. Mienert, F. Neese, K. Wieghardt, Science **312**, 1937 (2006)
5. N. Aliaga-Alcalde, S. DeBeer George, B. Mienert, E. Bill, K. Wieghardt, F. Neese, Angew. Chem. Int. Ed., 44, 2908 (2005)
6. M.P. Mehn, J.C. Peters, J. Inorg. Biochem. **100**, 634 (2006)
7. K. Meyer, J. Bendix, N. Metzler-Nolte, T. Weyhermüller, K. Wieghardt, J. Am. Chem. Soc. **120**, 7260 (1998)
8. K. Meyer, E. Bill, B. Mienert, T. Weyhermüller, K. Wieghardt, J. Am. Chem. Soc. **121**, 4859 (1999)
9. M. Schlangen, J. Neugebauer, M. Reiher, D. Schröder, J.P. López, M. Haryono, F.W. Heinemann, A. Grohmann, H. Schwarz, J. Am. Chem. Soc. **130**, 4285 (2008)
10. C.A. Grapperhaus, B. Mienert, E. Bill, T. Weyhermüller, K. Wieghardt, Inorg. Chem. **39**, 5306 (2000)
11. T.A. Betley, J.C. Peters, J. Am. Chem. Soc. **126**, 6252 (2004)
12. J.-U. Rohde, T.A. Betley, T.A. Jackson, C.T. Saouma, J.C. Peters, L. Que, Inorg. Chem. **46**, 5720 (2007)
13. J.-U. Rohde, J.-H. In, M.H. Lim, W.W. Brennessel, M.R. Bukowski, A. Stubna, E. Münck, W. Nam, L. Que, Science **299**, 1037 (2003)
14. X. Hu, K. Meyer, J. Organomet. Chem. **690**, 5474 (2005)
15. A.J. Arduengo III, PCT Int. Appl. WO 9117983 (du Pont de Nemours, E. I., and Co., USA) A1, 18 (1991)
16. K. Ward Jr., J. Am. Chem. Soc. **57**, 914 (1935)
17. N.N. Greenwood, T.C. Gibbs, *Mössbauer Spectroscopy* (Chapman and Hall Ltd., London, 1971)
18. http://ewww.mpi-muelheim.mpg.de/bac/logins/bill/julX_en.php
19. K. Koch, K. Koepernik, D.V. Neck, H. Rosner, S. Cottenier, Phys. Rev. A, 81, 032507 (2010)
20. P. Gütlich, Chemie in unserer Zeit **5**, 131 (1971)
21. P. Gütlich, Chemie in unserer Zeit **4**, 133 (1970)
22. I. Castro-Rodríguez, H. Nakai, K. Meyer, Angew. Chem. **118**, 2449 (2006)
23. T. Birk, J. Bendix, Inorg. Chem. **42**, 7608 (2003)
24. J. Bendix, R.J. Deeth, T. Weyhermüller, E. Bill, K. Wieghardt, Inorg. Chem. **39**, 930 (2000)
25. J. Bendix, K. Meyer, T. Weyhermüller, E. Bill, N. Metzler-Nolte, K. Wieghardt, Inorg. Chem. **37**, 1767 (1998)
26. J. Bendix, J. Am. Chem. Soc. **125**, 13348 (2003)
27. X. Hu, K. Meyer, J. Am. Chem. Soc. **126**, 16322 (2004)
28. R.O. Lindsay, C.F.H. Allen, Org. Synth. **22**, 96 (1942)
29. F. Neese, QCPE Bull. **15**, 5 (1995)

References

30. F. Neese, ORCA—An Ab Initio, Density Functional and Semiempirical Program Package, version 2.6.04; Institut für Physikalische und Theoretische Chemie, Universität Bonn, Germany (2007)
31. A.D. Becke, J. Chem. Phys. **98**, 5648 (1993)
32. C.T. Lee, W.T. Yang, R.G. Parr, Phys. Rev. B **37**, 785 (1988)
33. P.J. Stephens, F.J. Devlin, C.F. Chabalowski, M.J. Frisch, J. Phys. Chem. **98**, 11623 (1994)
34. A. Schäfer, H. Horn, R. Ahlrichs, J. Chem. Phys. **97**, 2571 (1992)
35. A. Schäfer, C. Huber, R. Ahlrichs, J. Chem. Phys. **100**, 5829 (1994)
36. SADABS 2.06 (Bruker AXS Inc., Madison, 2002)
37. SHELXTL NT 6.12 (Bruker AXS Inc., Madison, 2002)
38. C. Vogel, F.W. Heinemann, J. Sutter, C. Anthon, K. Meyer, Angewandte Chemie, Int Ed **47**, 2681 (2008)

Chapter 3
TIMEN$^{tol/3,5xyl}$: Unexpected Reactivity Resulting From Modifications of the Ligand Periphery

3.1 Introduction

Coordinatively unsaturated, electron-rich metal centers have proven to be powerful species for small molecule activation and functionalization [1]. Due to their σ-donor and π-accepting properties [2, 3], N-heterocyclic carbene (NHC) ligands are particularly suitable for the synthesis of a variety of low- to high-valent metal complexes; thus, they are perfectly suitable to act as supporting ligands for small molecule activation in reactive coordination complexes [4]. The steric bulk of NHC ligands is highly tunable and a variety of ligands can be conveniently synthesized. The anchoring unit of polydentate NHC systems provides the chelators with additional electronic and structural flexibility. It has been shown that sterically encumbering tripodal ligands of *tris*-[2-(3-*aryl*-**im**idazol-2-ylidene)ethyl]amine [TIMEN^R, with R = aryl = 2,6-xylyl (xyl), mesityl (mes)] create a trigonal platform for metal ions that enables the coordination and activation of small molecules of industrial and biological importance, such as dinitrogen, in a protective cylindrical cavity [5]. An active iron center with an iron nitride moiety has recently been suggested to be present at the site of biological nitrogen reduction [6] and significant progress in the synthesis and spectroscopic elucidation of iron imide, Fe=NR, and nitride species, Fe≡N, was achieved [7]. The first X-ray crystal structure determination of a discrete iron nitride complex, [(TIMEN^R)Fe(N)]$^+$, was accomplished after photolysis of an iron-azide complex stabilized by the N-anchored *tris*(carbene) ligand TIMEN^R (R = 2,6-xylyl, mesityl) [8]. Shortly after, the second X-ray crystallographic characterization of an iron nitride complex with a closely related borate-anchored *tris*(carbene) ligand was reported by Smith et al., namely [(PhB^{tBu}Im)Fe(N)], with (PhB^{tBu}Im)$^-$ = phenyl*tris*(1-*tert*-butylimidazol-2-ylidene)borato$^-$ [9], reflecting the potential of carbene based ligand systems, for the

---

C. S. Vogel, *High- and Low-Valent* tris-N-Heterocyclic Carbene Iron Complexes,
Springer Theses, DOI: 10.1007/978-3-642-27254-7_3,
© Springer-Verlag Berlin Heidelberg 2012

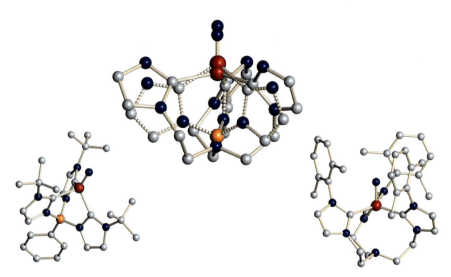

Fig. 3.1 Differences in the geometric structure of [(PhB'^BuIm)Fe(N)] and [(TIMEN^R)Fe(N)]^+

stabilization as well as for the synthesis of reactive iron nitride complexes. Interestingly, the reactivity of the N-anchored Fe≡N appears to be fundamentally different compared to the B-anchored system. While the B-anchored Fe≡N system shows reactivity towards a number of substrates, like the triphenylmethyl radical or TEMPO-H (1-hydroxy-2,2,6,6-tetramethyl-piperidine), the N-anchored Fe≡N complexes are remarkably inert towards the same substrates [10]. This difference in reactivity might be a result of the individual electronic and geometric structures of these two Fe(IV) systems. While the N-anchored complex is best described as trigonal pyramidal, the iron ion in the B-based Fe≡N is tetrahedrally coordinated (Fig. 3.1). Additionally, the steric demand of the two NHC tripods is markedly different. In contrast to the relatively accessible Fe≡N unit in [(PhB'^BuIm)Fe(N)], the nitride functionality in [(TIMEN^R)Fe(N)]^+ is situated in a deep cylindrical cavity with no side-access for substrates.

In order to synthesize N-anchored Fe≡N complexes with increased reactivity, a modification of the TIMEN^R ligand system by varying the steric demand of the aryl substituents was aimed for. In this chapter the synthesis and characterization of two new N-anchored NHC ligands of the TIMEN^R system, namely the tolyl and the 3,5-xylyl derivatived TIMEN^tol and TIMEN^{3,5xyl} is presented. During the synthesis of the corresponding iron complexes, unexpected reactions were observed resulting in a series of new four-, five-, and six-coordinate iron complexes. The synthesis, spectroscopic, and X-ray structural characterization of the iron complexes [(TIMEN^tol)Fe](BF_4)_2 (**3**), [(TIMEN^{tol***})Fe] (**4**), [(TIMEN^{3,5xyl})Fe(CH_3CN)](PF_6)_2 (**7**), [(TIMEN^{3,5xyl**})Fe] (**8**) and [(TIMEN^{3,5xyl**})Fe](PF_6) (**9**) is discussed (Chart 3.1). The electronic structures of **4**, **8**, and **9** were examined by density functional theory (DFT) calculations and are presented as well [11].

3.2 Results and Discussion

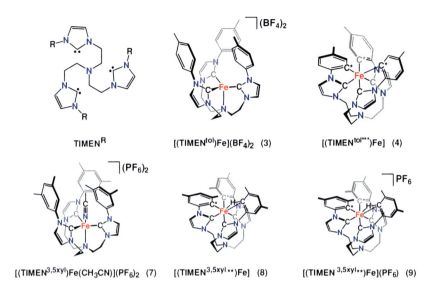

Chart 3.1 Overview of the tripodal TIMEN^R ligand (R = tolyl, 3,5-xylyl) and its newly synthesized iron complexes

3.2 Results and Discussion

The tripodal ligand system TIMEN^R (R = mesityl and 2,6-xylyl) has been successfully employed for the stabilization of metal centers in high and low oxidation states (see Chap. 2). As a remarkable result the synthesis and complete characterization of [(TIMEN^R)Fe(N)]BPh$_4$ was achieved [8]. This N-anchored Fe≡N complex is relatively unreactive and air- as well as moisture-stable in the solid state. Interestingly, preliminary attempts to oxidize the Fe(IV) nitride complexes to Fe(V) species, unexpectedly led to reduced *bis*-carbene imine species that formed via insertion of the terminal nitride ligand into one of the iron–carbene bonds of the TIMEN^R ligand (see Chap. 2). This chemistry is reminiscent of the insertion chemistry observed for the Co(III) imide complex, [(TIMEN^R)Co(NR)]BPh$_4$ [12]. The latter complex can be isolated at −35 °C but at room temperature it forms similar *bis*-carbene imine species in solution. Even in the presence of substrates, these examples of intramolecular insertion chemistry showed that the methyl groups in *ortho* position of the mesityl and 2,6-xylyl aryl rings of the NHC ligand play an important role in the reactivity of the corresponding iron complexes.

Due to a strong steric shielding of the substituents on these positions, side-access of potential substrates to the axial M≡L functional group is efficiently suppressed, thereby preventing further reactivity of the Fe≡N unit in [(TIMEN^R)Fe(N)]BPh$_4$ and its oxidized Fe(V) species (Fig. 3.2).

Aiming at high-valent metal complexes with axially bound terminal π-donor ligands, such as nitride, imides, and oxo ligands for atom and group transfer catalysis,

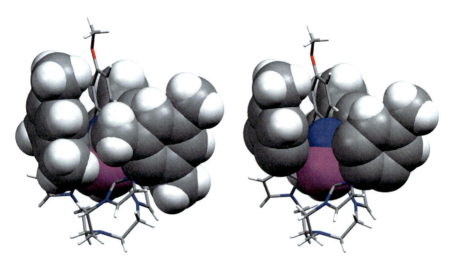

Fig. 3.2 Open side access by removal of 2,6-ortho-methyls in the mesityl and xylyl TIMEN[R] derivatives

two new tripodal carbene ligands with sterically less demanding aryl substituents, namely TIMEN[R] (R = aryl = tolyl (tol), 3,5-xylyl (3,5xyl), Scheme 3.1) were synthesized. These ligands were employed in the synthesis of the corresponding Fe(II) precursor complexes in order to investigate their potential for the preparation of new low- and high-valent iron complexes with enhanced reactivity.

Scheme 3.1 shows the general synthetic route to the tripodal ligand system. The first step is a one-pot reaction employing glyoxal, formaldehyde, ammonium salt, and an amine producing the derivatized imidazole [13]. The imidazoles can be linked to the framework anchor in an S_N^2-type reaction with *tris*(2-chloroethyl)amine [14]. Exchange of the Cl$^-$ counterions to PF$_6^-$ or BF$_4^-$ anions leads to non-hygroscopic imidazolium salts [15]. Deprotonation of the imidazolium salt with a strong base, like potassium *tert*-butoxide, yields the N-heterocyclic carbene. Reaction of the free carbene with a ferrous salt like FeCl$_2$ or FeOTf$_2$ gives the Fe(II) precursor complexes.

The cationic Fe(II) precursor [(TIMEN[tol])Fe]$^{2+}$ was obtained via a different but more convenient one-pot synthesis, a new and viable alternative route for the synthesis of Fe(II) precursors of the TIMEN[R] system. In this one-pot synthesis, a mixture of [H$_3$TIMEN[tol]](BF$_4$)$_3$, NaO^{t}Bu, NaBPh$_4$, and FeCl$_2$ in THF is stirred overnight, followed by recrystallization of the crude product. The advantage of the one-pot synthesis is that there is no need to isolate the free carbene, which is time consuming and requires a higher solvent consumption. Another point is that the iron(II) complex with tolyl substituent is difficult to isolate using the multi-step synthesis. Whereas the mesityl and 2,6-xylyl derivatives precipitate as analytical pure powder, from a mixture of ferrous chloride and carbene in pyridine, the tolyl Fe(II) complex precipitates under same conditions as a sticky mass, which is difficult to analyse because of poor solubility. The use of THF instead of pyridine

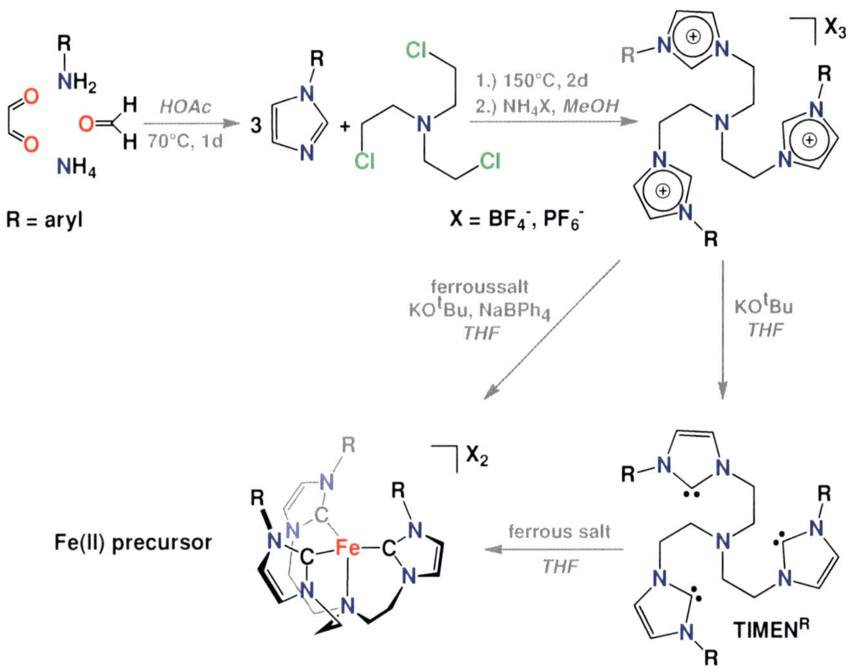

3.2 Results and Discussion

Scheme 3.1 Synthesis of TIMEN^R and the Fe(II) precursor complexes

improves the consistency, but the NMR reveals that beside the desired paramagnetic complex also the carbene precursor, [H$_3$TIMEN^{tol}]$^{3+}$ is formed again. Most probably also in this synthesis the remaining *tert*-butanol causes the problem. If the one-pot synthesis is performed without NaBPh$_4$, [H$_3$TIMEN^{tol}]$^{3+}$ is formed as well. Most likely, NaBPh$_4$ supports the dissolution of the different starting materials, like ferrous chloride and the imidazolium salt, thereby accelerating the coordination of the iron ion by the free carbene before it can be protonated again by *tert*-butanol, which is even more acidic in the presence of ferrous salt. The problem of the one-pot synthesis is, that the Fe(II) complex is contaminated with inorganic salt. But the crude product can be purified by recrystallization, which can be achieved by dissolving the crude product in acetonitrile and layering with diethyl ether at room temperature to obtain [(TIMEN^{tol})Fe](BF$_4$)$_2$ (**3**) analytically pure in about 50% yield.

The ^{1}H-NMR spectrum of complex **3** in solution is consistent with a paramagnetic high-spin iron(II) complex (Fig. 3.3). It shows five paramagnetically shifted resonances in the range between -5 and 55 ppm. Complex **3** exhibits a Mößbauer isomer shift, δ, of 0.51(1) mm s^{-1} ($\Delta E_Q = 1.54(1)$ mm s^{-1}, Fig. 3.4), slightly smaller than typically observed for four-coordinate Fe(II) high-spin complexes of this system ($\delta \approx 0.65$–0.75 mm s^{-1}) [8].

The molecular structure of the cation [(TIMEN^{tol})Fe]$^{2+}$ (Fig. 3.5) in crystals of [(TIMEN^{tol})Fe](BF$_4$)$_2$ · 0.8 Et$_2$O (**3** · 0.8 Et$_2$O), obtained from a solvent mixture

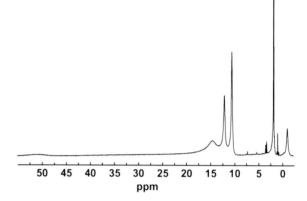

Fig. 3.3 ^{1}H-NMR spectrum of [(TIMEN^{tol})Fe](BF$_4$)$_2$ (3) (recorded in DMSO-d_6 at RT)

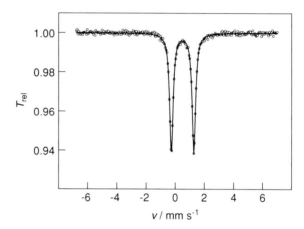

Fig. 3.4 Zero-field Mößbauer spectrum of [(TIMEN^{tol})Fe](BF$_4$)$_2$ (3) [recorded in solid state at 77 K; $\delta = 0.51(1)$ mm s^{-1}, $\Delta E_Q = 1.54(1)$ mm s^{-1}, $\Gamma_{FWHM} = 0.28(1)$ mm s^{-1}]

of acetonitrile and diethyl ether, exhibits a four-coordinate iron center with three equivalent iron carbene distances (average Fe–C$_{carbene}$ distance 2.051 Å) and an Fe–N$_{anchor}$ distance of 2.204(2) Å, clearly indicating a coordinated N-anchor. The metal center is located 0.197(2) Å below the trigonal plane of the three NHC carbene carbon atoms. It is remarkable that, although present in the reaction mixture, no axially bound ligand, like chloride or acetonitrile, is observed in the crystallographically determined structure of **3**. This is in contrast to the Fe(II) complexes of the related TIMEN^{mes} and TIMEN^{xyl} system with deep cylindrical cavities formed by the aryl substituents. For these complexes, axial coordination is always observed in the presence of coordinating anions or solvents. A possible reason for this could be the lack of methyl groups in 2- and 6-position of the aryl substituents in complexes of [(TIMEN^{tol})Fe]$^{2+}$, allowing free rotation of the tolyl arms of the ligand, thereby preventing the formation of an accessible ligand cavity for axial coordination in the TIMEN^{tol} system. This rotational degree of freedom is efficiently suppressed in ligands of the mesityl and 2,6-xylyl ligand derivatives.

3.2 Results and Discussion

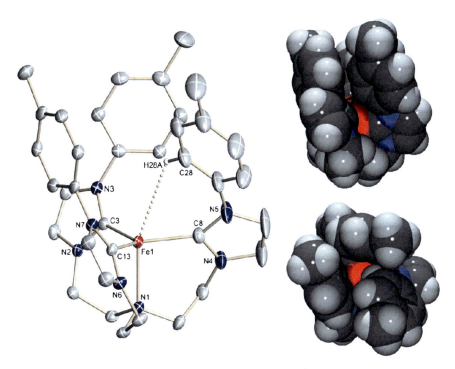

Fig. 3.5 Molecular structure of the complex cation of [(TIMEN^{tol})Fe](BF$_4$)$_2$ (**3**) in crystals of **3** · 0.8 Et$_2$O (50% probability ellipsoids). Hydrogen atoms (except H28A), BF$_4^-$ anions, and co-crystallized solvents are omitted for clarity. The *dotted line* indicates the weak anagostic Fe···H interaction. Space-filling representations of the cation **3** *side view* (*right, top*) and *top view* (*right, bottom*). Selected bond distances and angles (Å, °): Fe1–C3 2.046(2), Fe1–C8 2.055(2), Fe1–C13 2.052(2), Fe1–N1 2.204(2), Fe1···H28A 2.90, $d_{Fe\ oop}$ 0.197(2), Fe1···H28A–C28 122

Furthermore, close inspection of the structurally determined tolyl hydrogen atoms reveals that at least one of the *ortho* aryl C–H hydrogen atoms is pointing towards the center of the cavity and is thereby additionally blocking ligand access. The Fe1···H28A distance of 2.90 Å and the corresponding Fe1···H28A–C28 angle of 122° is indicative for a weak and largely electrostatic interaction (anagostic interaction) [16].

Attempts to reduce the Fe(II) ion in complex **3** with sodium amalgam yield a different product as compared with the mesityl and 2,6-xylyl derivatives (Scheme 3.2). Instead of the expected formation of the reduced Fe(I) species, the oxidized, threefold cyclometallated -aryl Fe(III) complex [(TIMEN^{tol***})Fe] (**4**) was isolated.

As expected for a low spin Fe(III) complex, the Mößbauer isomer shift of $\delta = 0.03(1)$ mm s^{-1} ($\Delta E_Q = 2.37(1)$ mm s^{-1}, Fig. 3.6) is considerably smaller than that of the Fe(II) precursor complex **3** (0.51 mm s^{-1}) and larger than that of the Fe(IV) nitride species (d^4, low-spin, $S = 0$, $\delta = -0.27$ mm s^{-1}). This confirms, that, despite the different coordination geometries and spin states of the

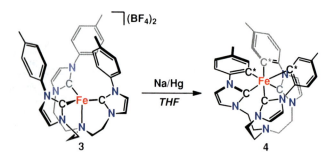

Scheme 3.2 Cyclometallation of [(TIMEN^{tol})Fe]$^{2+}$ (**3**) under reductive conditions

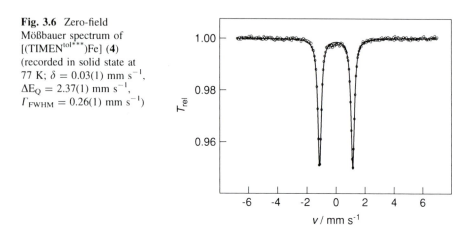

Fig. 3.6 Zero-field Mößbauer spectrum of [(TIMEN^{tol***})Fe] (**4**) (recorded in solid state at 77 K; $\delta = 0.03(1)$ mm s^{-1}, $\Delta E_Q = 2.37(1)$ mm s^{-1}, $\Gamma_{FWHM} = 0.26(1)$ mm s^{-1})

complexes, the trend of decreasing isomer shifts with increasing oxidation states in ^{57}Fe Mößbauer spectroscopy is confirmed within this series of complexes.

The molecular structure of [(TIMEN^{tol***})Fe] (**4**) (Fig. 3.7) in dark red crystals of [(TIMEN^{tol***})Fe] · 2 THF (**4** · 2 THF), obtained from a THF solution of the complex at −35 °C shows a pseudo-octahedral iron center coordinated by three NHC donors and three *ó*-aryl carbanions (denoted with an asterisk * in Scheme 3.2) with a *facial* arrangement of each of the two different sets of C donor atoms. In this complex, the iron center is coordinated by three bidentate ligand arms with every NHC carbon situated *trans* to a C* carbon atom of an aryl ring. It is remarkable, however, that in **4** one of the Fe–C* bonds, Fe–C24 = 2.366(2) Å, is significantly longer, by approximately 0.35 Å, than the other two Fe–C* bonds. The shortest Fe–C$_{carbene}$ bond (Fe–C3 = 1.888(2) Å) is *trans* to this longest bond (Table 3.1). In contrast, a perfectly symmetrical arrangement is observed for *fac*-[Ir{CN(C$_6$H$_4$Me-*p*)(CH$_2$)$_2$NC$_6$H$_3$Me-*p*}$_3$] [17], a homoleptic complex exhibiting exactly the same coordination mode with three NHC donors (Ir–C$_{avg}$ 2.03(3) Å) and three *trans* coordinated aryl carbanions (Ir–C$_{avg}$ 2.09(3) Å). It is reasonable to assume that the three individual bidentate ligands, coordinating the iridium in

3.2 Results and Discussion 61

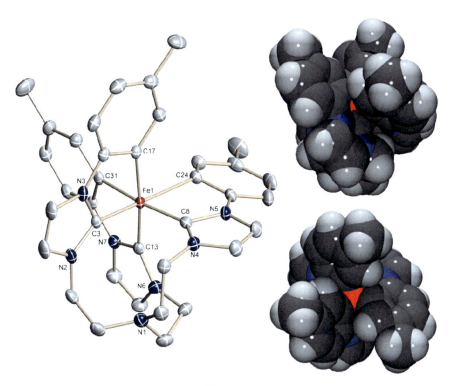

Fig. 3.7 Molecular structure of [(TIMEN^{tol***})Fe] (**4**) in crystals of **4** · 2 THF (50% probability ellipsoids). Hydrogen atoms and co-crystallized solvents omitted for clarity. Space-filling representations of the cation **4** *side view* (*right, top*) and *top view* (*right, bottom*). Selected bond distances (Å): Fe1–C3 1.888(2), Fe1–C8 1.923(2), Fe1–C13 1.943(2), Fe1–N1 3.947(2), Fe1–C17 1.994(2), Fe1–C24 2.366(2), Fe1–C31 2.014(2)

fac-[Ir{CN(C$_6$H$_4$Me-*p*)(CH$_2$)$_2$NC$_6$H$_3$Me-*p*}$_3$], are free to arrange in a symmetrical fashion around the metal center, while the flexibility of the tripodal TIMEN^R ligand with three N-tethered carbene aryl entities is reduced, resulting in a non-symmetrical binding mode to the smaller iron ion. This may also result in ligand distortions. For instance, in two of the ligand arms in **4** the five-membered NHC ring and the adjacent aryl ring are almost coplanar, whereas the two rings of the third ligand arm with a considerably longer Fe–C* bond exhibit a significant deviation from coplanarity. The torsion angles C*–C$_{aryl}$–N–C$_{carbene}$ are a good measure of the coplanarity of the five-membered imidazol-2-ylidene and aryl ring systems and amount to only 0.6(2)° for C3–N3–C16–C17 and 2.4(3)° for C13–N7–C30–C31, but to 17.1(3)° for C8–N5–C23–C24. The EPR spectrum of a frozen solution of **4** shows an axial-symmetrical signal with resonances centered at $g_{\|} = 2.29$ and $g_{\perp} = 1.94$ (Fig. 3.8). This is in agreement with the complexes' axially distorted octahedral symmetry and a ground state of $S = 1/2$.

In an effort to suppress the observed intramolecular metallacycle formation, an alternative approach was attempted. The introduction of 3,5-xylyl substituents

Table 3.1 Selected bond distances (Å) and angles (°) for molecular structures of **3**, **4**, **7**, **8** and **9** (e.s.d's in parentheses)

Compounds	3	4	7	8	9
Fe–C3	2.046(2)	1.888(2)	2.101(3)	1.913(2)	1.972(3)
Fe–C8	2.055(2)	1.923(2)	2.108(3)	1.870(2)	1.895(3)
Fe–C13	2.052(2)	1.943(2)	2.107(3)	1.941(2)	1.999(3)
Fe–N1	2.204(2)	3.947(2)	2.545(3)	3.936(2)	3.935(3)
Fe–N8	–	–	2.193(2)	–	–
Fe–C17	–	1.994(2)	–	2.442(2)	2.525(3)
Fe–C24[a]	–	2.366(2)	–	2.050(2)	2.032(3)
Fe–C31[b]	–	2.014(2)	–	2.074(2)	2.037(3)
C3–Fe–C8	121.73(9)	100.01(8)	117.9(2)	101.10(8)	103.4(2)
C3–Fe–C13	118.81(8)	97.59(9)	120.1(2)	93.73(8)	91.3(2)
C8–Fe–C13	116.73(9)	97.46(9)	120.2(2)	95.53(8)	93.9(2)
C3–Fe–N1	96.67(8)	–	85.0(1)	–	–
C8–Fe–N1	95.34(8)	–	85.6(1)	–	–
C13–Fe–N1	94.45(8)	–	86.0(1)	–	–
C3–Fe–C17	–	82.12(9)	–	78.52(8)	
C8–Fe–C17	–	89.77(8)	–	176.17(7)	
C13–Fe–C17	–	172.69(8)	–	80.71(8)	
C3–Fe–C24[a]	–	178.01(8)	–	87.40(8)	86.6(2)
C8–Fe–C24[a]	–	78.86(8)	–	82.53(8)	82.1(2)
C13–Fe–C24[a]	–	84.20(8)	–	177.92(8)	174.9(2)
C3–Fe–C31[b]	–	89.12(8)	–	169.64(8)	169.1(2)
C8–Fe–C31[b]	–	170.87(8)	–	88.44(8)	85.9(2)
C13–Fe–C31[b]	–	91.69(8)	–	81.22(8)	82.2(2)
$d_{Fe\ oop}$	0.197(2)	–	0.164(2)	–	–

[a] C24 corresponds to C25 in complexes **8** and **9**
[b] C31 corresponds to C33 in complexes **8** and **9**

into the TIMEN-ligand system (instead of mesityl and 2-6-xylyl groups) is expected to slightly increase the distance of the *ortho* C–H groups from the Fe center. This should effectively prevent metallation of the reactive *ortho* C–H bonds, and thus, provide a chelating ligand with increased side-access to the $M \equiv E$ functional group.

Treatment of the free 3,5-xylyl derivatized *tris*(carbene) with ferrous chloride in a mixture of THF and pyridine, followed by recrystallization from acetonitrile, yields the corresponding Fe(II) precursor complex [(TIMEN^{3,5xyl})Fe(CH$_3$CN)](PF$_6$)$_2$ (**7**) (Fig. 3.11). The ^{1}H-NMR spectrum of complex **7** in solution is consistent with a paramagnetic high-spin iron(II) complex (Fig. 3.10). It shows several paramagnetically shifted resonances, partially extremely broadend, in the range between 0 and 30 ppm, which are difficult to assign. Complex **7** exhibits a Mößbauer isomer shift of $\delta = 0.73(1)$ mm s^{-1} ($\Delta E_Q = 1.08(1)$ mm s^{-1}, Fig. 3.9), characteristic of a four-coordinate Fe(II) high-spin ($S = 2$) species of this system.

3.2 Results and Discussion

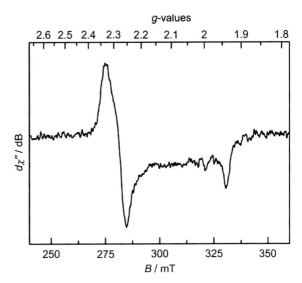

Fig. 3.8 X-band EPR spectrum of [(TIMEN^{tol***})Fe] (**4**) recorded in frozen acetonitrile solution at 7.1 K, $v = 8.9875$ GHz, P = 1.0 mW, $g_{\parallel} = 2.29$, $g_{\perp} = 1.94$

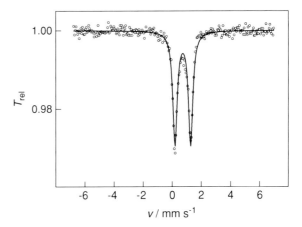

Fig. 3.9 Zero-field Mößbauer spectrum of [(TIMEN^{3,5xyl})Fe(CH$_3$CN)](PF$_6$)$_2$ (**7**) (recorded in solid state at 77 K; $\delta = 0.73(1)$ mm s^{-1}, $\Delta E_Q = 1.08(1)$ mm s^{-1}, $\Gamma_{FWHM} = 0.26(1)$ mm s^{-1})

In contrast to [(TIMEN^{tol})Fe]$^{2+}$ (**3**) with rotatable tolyl substituents, rotation of the aryl substituents in **7** is hindered by the methyl groups in 3,5-position. Therefore, the steric demand of the 3,5-substituted aryl arm in **7** is higher and results in an axial cavity large enough to accommodate an axially bound acetonitrile ligand. The Fe(II) center in crystals of [(TIMEN^{3,5xyl})Fe(CH$_3$CN)](PF$_6$)$_2$ · 2 CH$_3$CN (**7** · 2 CH$_3$CN), obtained by slow diethyl ether diffusion into a layered acetonitrile solution of the complex, is coordinated in a trigonal pyramidal fashion with three equivalent Fe–NHC carbon bonds with an average Fe–C$_{carbene}$ distance of 2.105 Å and a coordinated acetonitrile ligand (Fe–N$_{CH3CN}$ distance of 2.193(2) Å). The iron center is situated 0.164(2) Å above the trigonal plane formed by the three NHC carbene donor atoms. Due to the axial coordination of an acetonitrile molecule, the Fe–N$_{anchor}$ distance of 2.545(3) Å is considerably longer

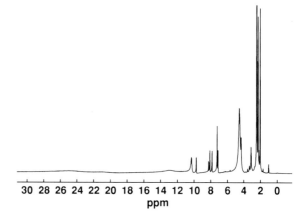

Fig. 3.10 ¹H-NMR spectrum of [(TIMEN³,⁵ˣʸˡ)Fe(CH₃CN)](PF₆)₂ (**7**) (recorded in DMSO-d_6 at RT)

than that found in [(TIMEN^tol)Fe]²⁺ and shows that the N-anchor is no longer coordinated in **7** (Fig. 3.11).

The ability of the [(TIMEN³,⁵ˣʸˡ)Fe]²⁺ moiety to accommodate an axial ligand (as was observed earlier for the corresponding mesityl and 2,6-xylyl derivatives) prompted us to attempt the synthesis of a high-valent iron nitride complex. In order to synthesize the required Fe(I) precursor complexes [(TIMEN^R)Fe]⁺, reduction of **7** using the sodium amalgam protocol used for the reduction of [(TIMEN^mes/xyl)Fe(Cl)]⁺ was performed. However, in contrast to the complexes of the mesityl- and 2,6-xylyl ligand derivatives, and reminiscent of the observations made for the tolyl derivative **3** (vide supra), the reduction of the Fe(II) complex **7** yielded yet another Fe(II) metallation product (Scheme 3.3).

The Mößbauer spectrum of a solid sample of this reaction shows two doublets (Fig. 3.12). The doublet with an isomer shift, δ, of 0.07(1) mm s⁻¹ and a quadrupole splitting, ΔE_Q, of 2.37(2) mm s⁻¹ belongs to the Fe(II) complex [(TIMEN³,⁵ˣʸˡ**)Fe] (**8**), whereas the second doublet with an isomer shift, γ, of 0.88(1) mm s⁻¹ and a quadrupole splitting, ΔE_Q, of 2.57(2) mm s⁻¹ belongs to the byproduct. From the Mößbauer parameters obtained for this byproduct it can be assumed that it could be an Fe(I) species, for instance the corresponding *mono*-metallated iron complex. However attempts to isolate and characterize this byproduct remained unsuccessful. The Mößbauer parameters obtained for [(TIMEN³,⁵ˣʸˡ**)Fe] (**8**) are remarkably different from its Fe(II) precursor ($\gamma = 0.73(1)$ mm s⁻¹, $\Delta E_Q = 1.08(1)$ mm s⁻¹), and are indicative of an Fe(II) low spin ($S = 0$) species in an octahedral ligand environment, as opposed to the four-coordinate Fe(II) high-spin ($S = 2$) precursor. However, the metallated Fe(II) complex does not give ¹H-NMR spectra, suggesting that this complex is a high spin, $S = 2$, in solution.

The results of an X-ray crystal structure determination confirmed the formation of a pseudo-octahedrally coordinated *bis*-metallated Fe(II) complex, [(TIMEN³,⁵ˣʸˡ**)Fe] (**8**) (Fig. 3.13), rather than the *tris*-metallated Fe(III) species that was obtained in case of the tolyl derivative (compound **4**).

3.2 Results and Discussion

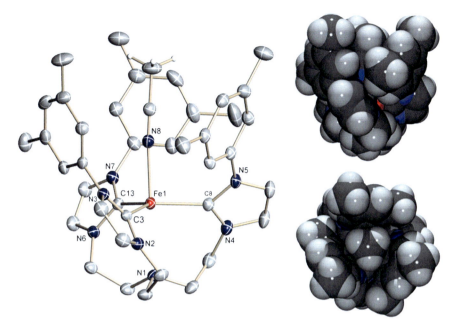

Fig. 3.11 Molecular structure of the complex cation of [(TIMEN^{3,5xyl})Fe(CH$_3$CN)] (PF$_6$)$_2$ (**7**) in crystals of **7** · 2 CH$_3$CN (50% probability ellipsoids). Hydrogen atoms of the TIMEN^{3,5xyl} ligand, the PF$_6^-$ anions, and co-crystallized solvents are omitted for clarity. Space-filling representations of the cation **7** *side view* (*right, top*) and *top view* (*right, bottom*). Selected bond distances (Å): Fe1–C3 2.101(3), Fe1–C8 2.108(3), Fe1–C13 2.107(3), Fe1–N1 2.545(3), Fe1–N8 2.193(2), $d_{Fe\ oop}$ 0.164(2)

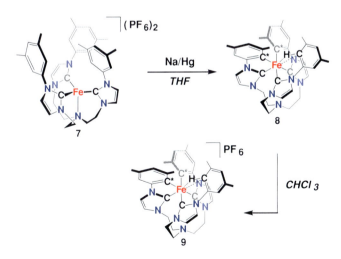

Scheme 3.3 Cyclometallation of [(TIMEN^{3,5xyl})Fe]$^{2+}$ under reductive conditions

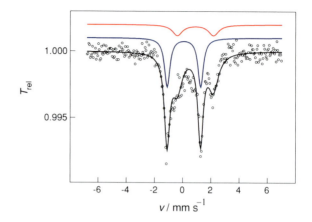

Fig. 3.12 Zero-field Mößbauer spectrum of [(TIMEN^{3,5xyl**})Fe] (**8**) [recorded in solid state at 77 K; $\delta = 0.07(1)$ mm s^{-1}, $\Delta E_Q = 2.37(1)$ mm s^{-1}, $\Gamma_{FWHM} = 0.47(1)$ mm s^{-1}; 57%, *blue line*. Byproduct: $\delta = 0.88(1)$ mm s^{-1}, $\Delta E_Q = 2.57(1)$ mm s^{-1}, $\Gamma_{FWHM} = 0.87(1)$ mm s^{-1}; 43%, *red line*]

The pseudo-octahedral iron center in **8** coordinates to three facially arranged N-heterocyclic carbene donors, two ó-aryl carbanions (denoted with an asterisk * in Scheme 3.3). The iron ion's distorted octahedral coordination environment is completed with an agostic hydrogen interaction to one of the aryl's *ortho* CH groups, namely H17A. The position of the hydrogen atom of this non-dehydrogenated aryl CH group could be determined experimentally from a difference Fourier map. However, for reasons of consistency, H17A was placed in an idealized position of optimized geometry during refinement. The Fe–C bond distances in **8** are similar to those observed in **4** and the shorter NHC–Fe bond distances are compensated by longer Fe–C* carbanion distances. The shortest iron carbene distance of 1.870(2) Å for Fe–C8 is observed *trans* to the CH of the non-dehydrogenated ring (Fe⋯C17 = 2.442(2) Å and Fe⋯H17A = 1.71 Å, Fe1⋯H17A–C17 = 131°), which also exhibits the strongest deviation from co-planarity of the NHC and the 3,5-xylyl rings. When comparing the distortion between the NHC and aryl moieties in both, the tolyl and the 3,5-xylyl derivatives, it is obvious that the 3,5-substitution of the aryl ring is increasing the steric demand. As observed in the molecular structure of **4**, two of the NHC–aryl systems of **8** are almost perfectly coplanar with corresponding torsion angles of 4.3(2)° for C8–N5–C24–C25 and 1.7(2)° for C13–N7–C32–C33. The N-heterocyclic carbene and 3,5-xylyl moieties of the third pendant ligand arm, however, are significantly tilted, reflected by the C3–N3–C16–C17 torsion angle of 27.4(3)° (as compared to 17.1(3)° found in the sterically less crowded tolyl derivative **4**).

The *bis*-metallated complex [(TIMEN^{3,5xyl**})Fe] (**8**) can easily be oxidized to yield a green colored Fe(III) complex (Scheme 3.3). This oxidation occurs readily even during work-up of **8** in halogenated solvents. A straightforward synthesis was accomplished by treatment of **8** with silver salts, like AgOTf, in acetonitrile. The molecular structure of the resulting complex [(TIMEN^{3,5xyl**})Fe](PF$_6$) (**9**) is depicted in Fig. 3.14.

In analogy to **4**, the EPR spectrum of a frozen acetonitrile solution of **9**, with g-values of $g_1 = 2.42$, $g_2 = 2.20$, $g_3 = 1.94$, confirms the $S = 1/2$ ground state of

3.2 Results and Discussion

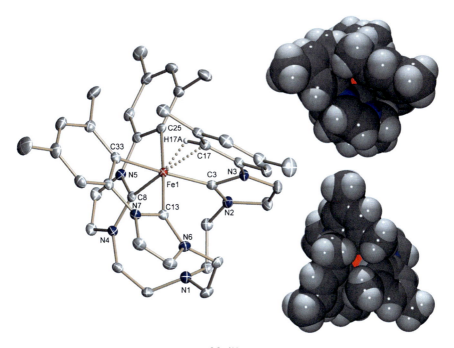

Fig. 3.13 Molecular structure of [(TIMEN^{3,5xyl**})Fe] (**8**) (50% probability ellipsoids). Hydrogen atoms (except H17A) omitted for clarity. *Dotted lines* indicate the agostic Fe⋯H interaction Fe1–N1 2.545(2). Space-filling representations of the cation **8** *side view* (*right*, *top*) and *top view* (*right*, *bottom*). Selected bond distances (Å): Fe1–C3 1.913(2), Fe1–C8 1.870(2), Fe1–C13 1.941(2), Fe1–N1 3.936(2), Fe1–C17 2.442(2), Fe1–C25 2.050(2), Fe1–C33 2.074(2)

this low-spin Fe(III) complex (Fig. 3.15). In **9**, the Fe(III) center is coordinated pseudo-octahedrally and all metric parameters of the molecular structure are very similar to those observed for the Fe(II) complex **8**. Slight differences are observed only for the Fe–C distances with longer iron carbene bonds and shorter iron carbanion bonds compared to those found in the Fe(II) species (Table 3.1). Compared to **8**, the Fe(III)–CH$_{aryl}$ bond distance in **9** is slightly longer and was determined to be Fe⋯C17 = 2.525(3) Å. The other relevant distances are Fe⋯H17A = 1.79 Å and Fe1⋯H17A–C17 = 131° and are again representing an agostic interaction. This elongation of bond distances in **9** is most likely due to the smaller ionic radius of the Fe(III) ion, compared to the divalent Fe center in **8**. The third arm of the ligand shows a significant tilt between the NHC and the 3,5-xylyl moiety shown by the C3–N3–C16–C17 torsion angle of 25.7(4)°. The other two NHC carbene aryl moieties remain coplanar (the corresponding torsion angles C8–N5–C24–C25 and C13–N7–C32–C33 amount to 0.1(4)° and 0.4(4)°, respectively). In general, the overall coordination environment with two almost ideally coplanar arranged aryl–carbene moieties, short iron NHC distances, and one elongated iron CH$_{aryl}$ distance, corresponding to the tilted ligand arm *trans* to the shortest Fe–NHC bond, does not change between the *bis*-metallated Fe(II) and Fe(III) centers in complexes **8** and **9**.

68 3 TIMEN^(tol/3,5xyl): Unexpected Reactivity Resulting From Modifications

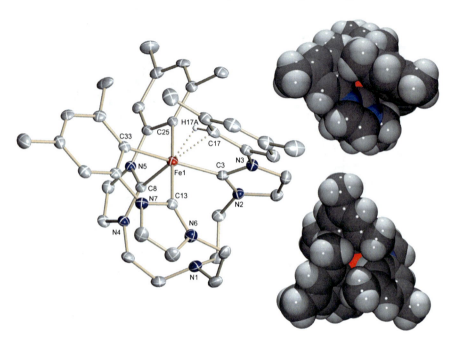

Fig. 3.14 Molecular structure of the complex cation of [(TIMEN^{3,5xyl**})Fe]$^+$ (**9**) in crystals of **9** · 0.6 THF · 0.4 CHCl$_3$ (50% probability ellipsoids). Hydrogen atoms (except H17A), PF$_6^-$ anion, and co-crystallized solvents are omitted for clarity. The *dotted lines* indicate the agostic Fe···H interaction. Space-filling representations of the cation **9** *side view* (*right, top*) and *top view* (*right, bottom*). Selected bond distances (Å): Fe1–C3 1.972(3), Fe1–C8 1.895(3), Fe1–C13 1.999(3), Fe1–N1 3.935(3), Fe1–C17 2.525(3), Fe1–C25 2.032(3), Fe1–C33 2.037(3)

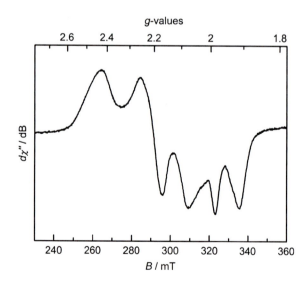

Fig. 3.15 X-band EPR spectrum of [(TIMEN^{3,5xyl**})Fe](PF$_6$) (**9**) recorded in frozen acetonitrile solution at 68 K, ν = 8.9817 GHz, P = 1.0 mW, g_1 = 2.42, g_2 = 2.20, g_3 = 1.94

Table 3.2 Principal bond distances of the optimized structures as obtained from the spin-unrestricted BP-DFT calculations

	4	8	9
Fe–C_{NHC}	1.964 (1.943)[a]	1.897 (1.913)	2.011 (1.990)
Fe–C*[b]	2.016 (1.994)	2.092 (2.074)	2.041 (2.032)
Fe–C_{NHC}	1.962 (1.923)	1.940 (1.940)	1.983 (1.972)
Fe–C*	2.014 (2.014)	2.074 (2.050)	2.055 (2.037)
Fe–C_{NHC}	1.962 (1.888)	1.872 (1.870)	1.882 (1.895)
Fe–C*	2.015 (2.366)	2.468 (2.442)	2.593 (2.524)
Fe···H[c]	–	1.782 (1.708)	1.883 (1.795)

[a] The experimental values are given in parentheses
[b] *trans* to C_{NHC}
[c] Agostic interactions

3.3 Theoretical Considerations

The electronic structures of **4**, **8**, and **9** were additionally examined by density functional theory (DFT) calculations. The geometries of the complexes were fully optimized using the BP86 functional (Table 3.2). The optimized structures of **8** and **9** are in excellent agreement with the available X-ray structures showing the presence of agostic Fe···H interactions (Tables 3.1 and 3.2). In contrast, the experimental structure of **4** possesses an axially distorted octahedral geometry, whereas the calculated structure is symmetrical, showing a set of three similar Fe–C* and a set of three similar Fe–C_{NHC} bonds.

Attempts to optimize the molecular structure of **4** using larger basis sets, which include relativistic effects, a conductor like screening model (COSMO), and empirical van der Waals corrections all gave very similar, symmetrical structures. Therefore, based on the DFT calculations, it appears that in principle the ligand is sufficiently flexible to coordinate the Fe ion in a symmetrical fashion with three facially coordinating NHC carbon and three metallated aryl carbon ligands. It is, however, quite remarkable that the *tris*-metallated Fe(III) complex **4** and the *bis*-metallated Fe(II) and Fe(III) complexes **8** and **9** show a similar departure from octahedral symmetry with an axial distortion (one short Fe–C_{NHC} and one long Fe–C*$_{aryl}$ bond and agostic Fe···CH interaction *trans* to it, respectively, see Fig. 3.16 and Table 3.3).

It can be speculated that the observed structural distortion in **4** originates from electronic (Jahn–Teller distortion), steric (ligand flexibility) or even crystal packing effects. The phenomenon appears too pronounced, however, to be entirely due to crystal packing effects. It also seems unlikely that such a packing effect can be observed in three different crystal structures. Alternatively, one could also argue that d^5 low spin complexes ($S = 1/2$), such as **4** and **9**, are subject to a Jahn–Teller distortion, and hence, the observed structural distortion could be the result of the Jahn–Teller effect, leading to a non-degenerate ground-state. However, complex **8** is a low spin Fe(II) complex ($S = 0$), for which no Jahn–Teller effect is expected.

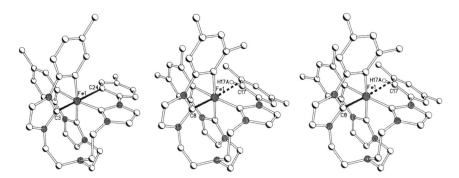

Fig. 3.16 Molecular representations of complexes **4** (*left*), **8** (*middle*), and cation **9** (*right*), the axial distortion highlighted in *bold*

Table 3.3 Principal bond distances of the observed structures as obtained from the X-ray crystal structure determination

	4	8	9
Fe–C$_{NHC}$	1.943[a]	1.913	1.999
Fe–C*	1.994	2.074	2.032
Fe–C$_{NHC}$	1.923	1.941	1.972
Fe–C*	2.014	2.050	2.037
Fe–C$_{NHC}$	1.888	1.870	1.895
Fe – C	2.366	–	–
Fe⋯CH[a]	–	2.053	2.138

[a] These values refer to the distance between Fe and the middle of the CH bond

On the other hand, while there is no apparent steric hindrance in **4**, it appears reasonable to assume that steric pressure, exerted by the aryl methyl groups in the 3,5-xylyl-derivatized species, effectively prevents *tris*-metallation in complex **8**, which results in one weak Fe⋯CH agostic interaction and one short Fe–C$_{NHC}$ bond of 1.870 Å *trans* to it. This may also explain the torsion angles observed within the five-membered NHC and the aryl substituent, which are more pronounced in the [(TIMEN^{3,5xyl**})Fe]$^{0/+}$ complexes **8** and **9** compared to the tolyl-derivatized species **4**. In summary, whereas the axially distorted, pseudo-octahedral structures of **8** and **9** can be theoretically described and explained on the basis of steric pressure, the reason for the axial distortion in **4** and its discrepancy to the computationally calculated structure remains unclear.

The Mößbauer parameters for **4**, **8**, and **9** were also calculated at the B3LYP level of theory (Table 3.4). Interestingly, and despite the failure to reproduce the experimentally determined structure of **4**, all calculated parameters reproduce the experimentally determined values accurately. The calculated complexes exhibit isomer shifts typical for low-spin ferrous and low-spin ferric ions. The calculated electronic structures confirm the presence of a low-spin ferrous ion in **8** and a low-spin ferric ion in **4** and **9**. Several attempts to calculate intermediate- and high-spin

3.3 Theoretical Considerations

Table 3.4 Mößbauer parameters of **4**, **8**, and **9** obtained from the spin-unrestricted B3LYP-DFT calculations

	δ (mm s^{-1})	ΔE_Q (mm s^{-1})
[(TIMEN^{tol***})Fe] (**4**)	-0.11 (+0.03)[a]	+2.12 (2.37)
[(TIMEN^{3,5xyl**})Fe] (**8**)	+0.05 (+0.07)	+2.34 (2.37)
[(TIMEN^{3,5xyl**})Fe](PF$_6$) (**9**)	-0.01[b]	1.30[b]

[a] The experimental values are given in parentheses
[b] No experimental data available

species resulted in significant discrepancy between the calculated and the experimental molecular structures. This further confirms the low-spin ground states for **4**, **8**, and **9**. All doubly and singly occupied molecular orbitals (MOs) with significant iron d-character were identified among the sets of canonical or quasi-restricted MOs (Figs. 3.17, 3.18 and 3.19). The unoccupied d-orbitals are highly delocalized and are not shown. The approximate C_3 symmetry of the complex **4** results in strong mixing of d-orbitals within the t_{2g} set. However, d-orbital mixing in **8** and **9** is lower than in **4** due to the less symmetrical molecular structures of **8** and **9**. Hence, the one-electron oxidation **8** → **9** can computationally be shown to result in the loss of one electron mainly from the d_{xy} metal orbital.

3.4 Conclusion

The exceptional stability, and thus, reduced reactivity of the first generation nitride [(TIMEN^{mes})Fe(N)]$^+$ was attributed to the steric demand exerted by the mesityl and 2,6-xylyl substituents. In an effort to enhance the reactivity of the various low- and high-valent iron complexes of the TIMEN^R ligand system [R = aryl = mesityl (mes), 2,6-xylyl (xyl)], two new ligands and their corresponding iron complexes were synthesized and spectroscopically characterized. Throughout the syntheses of the iron complexes, unexpected reactivities of the C–H bonds in ortho-positions of the newly employed aryl derivatives were observed with these new ligand systems, resulting in a series of new four, five, and six-coordinate complexes. The divalent iron precursor complexes **3** and **7**, *tris*-metallated **4**, and the *bis*-metallated complexes **8** and **9**, as well as the imidazolium salts of the ligands (see Experimental) were characterized by single crystal X-ray structure determination and spectroscopic methods.

As predicted, introduction of sterically less demanding aryl substituents bearing potential reactive C–H bonds in *ortho* position of the aromatic rings changes the reactivity of the Fe(II) precursor complexes. Introduction of the tolyl group reduced the steric pressure in a way that the aryl groups are now free to rotate, thereby preventing coordination of an axial ligand in the corresponding Fe(II) complex **3**. In contrast, replacing the tolyl with 3,5-xylyl groups resulted in the formation of an Fe(II) precursor complex **7** with an accessible axial coordination cavity. From the structure determinations of the Fe(II) precursor

72　　　　3　TIMEN$^{tol/3,5xyl}$: Unexpected Reactivity Resulting From Modifications

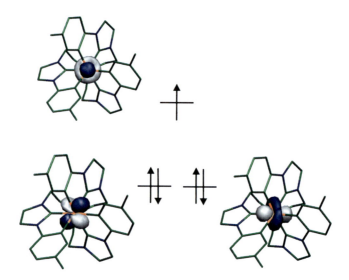

Fig. 3.17 Occupied iron-based frontier MOs for **4** as obtained from the spin-unrestricted B3LYP-DFT calculations: quasi-restricted orbitals are shown, all three MOs are mixtures of d_{xy}, d_{xz}, and d_{yz} iron orbitals

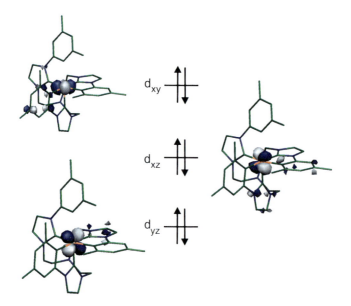

Fig. 3.18 Occupied iron-based frontier MOs for **8** as obtained from the spin-unrestricted B3LYP-DFT calculations: canonical orbitals are shown

complexes **3** and **7** it is evident that the steric demand is significantly reduced. However, these Fe(II) precursor complexes exhibit unexpected reactivities that

3.4 Conclusion

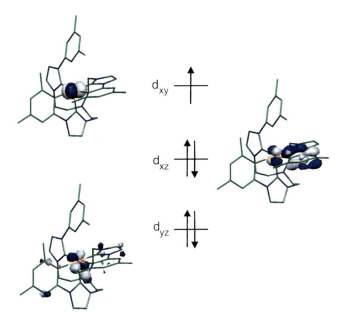

Fig. 3.19 Occupied iron-based frontier MOs for **9** as obtained from the spin-unrestricted B3LYP-DFT calculations: quasi-restricted orbitals are shown

differ significantly from the mesityl and 2,6-xylxl derivatives. Attempts to reduce the triganol planar and trigonal pyramidal Fe(II) precursors over sodium amalgam did not yield the corresponding Fe(I) complexes. Instead, cyclometallation of the aryl carbene moieties leading to pseudo-octahedrally coordinated, *tris*-metallated (in the case of the tolyl system) or *bis*-metallated (for the 3,5-xylyl system) Fe(II) and Fe(III) complexes. While the *tris*-metallated species [(TIMEN^{tol***})Fe] (**4**) is relatively stable towards oxidation, the *bis*-metallated [(TIMEN^{3,5xyl**})Fe] (**8**) is easily oxidized to yield the corresponding Fe(III) compound [(TIMEN^{3,5xyl**})Fe](PF$_6$) (**9**). The structural parameters demonstrate that this ligand framework is well suited to coordinate iron centers in a trigonal pyramidal fashion with three equivalent ligand arms. An octahedral coordination of the central metal requires a twist of the third ligand arm that features large twist angles between the five-membered NHC ring and the adjacent aryl ring of 17.1(3)° (for **4**), 27.4(3)° (for **8**) and of 25.7(4)° (for **9**). Reducing the steric demand of the TIMEN^R ligand system by introducing aryl rings with no substituents in the 2,6-position was successfully achieved. However, the two new ligands TIMEN^{tol} and TIMEN^{3,5xyl} yielded Fe(II) precursor complexes **3** and **7** that again showed unexpected reactivities upon reduction, very different to those observed for the mesityl and 2,6-xylyl derivatives. Obviously, relatively small changes in the periphery of the ligand have drastic effects on the reactivity of the resulting complexes. Further fine-tuning of the TIMEN^R ligand system by introducing other aryl substituents seems to be necessary to synthesize $Fe \equiv N$ complexes with enhanced reactivity.

3.5 Experimental

3.5.1 Methods, Procedures, and Starting Materials

All air- and moisture-sensitive experiments were performed under dry nitrogen atmosphere using standard Schlenk line techniques or an MBraun inert-gas glovebox containing an atmosphere of purified dinitrogen. Anhydrous iron(II) chloride, 99.9%, was purchased from Aldrich and used as received. Sodium tetraphenylborate, potassium *tert*-butoxide, and sodium *tert*-butoxide were purchased from ACROS and were used without further purification. Solvents were purified using a two-column solid-state purification system (Glasscontour System, Irvine, CA) and transferred to the glove box without exposure to air. NMR solvents were obtained packaged under argon and stored over activated molecular sieves and sodium (where appropriate) prior to use. *Tris*(2-chloroethyl)amine, 1-(3,5-xylyl)imidazole, and 1-(tolyl)imidazole were synthesized according to literature [13, 14]. Elemental analyses were performed at the Analytical Laboratories of the Friedrich-Alexander-University Erlangen-Nuremberg (Erlangen, Germany). Due to poor solubility, limited stability, and difficulties in separation of the reactants during work-up of some of the compounds, certain routine analytical methods could not be applied. Yields were calculated and noted where possible. All compounds were characterized by single crystal X-ray structure determination with crystals being representative for the bulk sample.

^{1}H-NMR spectra were recorded on JEOL 270 and 400 MHz instruments, operating at respective frequencies of 269.714 and 400.178 MHz with a probe temperature of 23 °C. ^{13}C-NMR spectra were recorded on JEOL 270 and 400 MHz instruments, operating at respective frequencies of 67.82 and 100.624 MHz with a probe temperature of 23 °C. Chemical shifts were reported in ppm relative to the peak of $SiMe_4$ using ^{1}H (residual) chemical shifts of the solvent as a secondary standard.

^{57}Fe Mößbauer spectra were recorded on a WissEl Mößbauer spectrometer (MRG-500) at 77 K in constant acceleration mode. ^{57}Co/Rh was used as the radiation source. WinNormos for Igor Pro software was used for the quantitative evaluation of the spectral parameters (least-squares fitting to Lorentzian peaks). The minimum experimental line widths were 0.23 mm s^{-1}. The temperature of the samples was controlled by an MBBC-HE0106 MÖSSBAUER He/N$_2$ cryostat within an accuracy of ± 0.3 K. Isomer shifts were determined relative to α-iron at 298 K.

EPR measurements were performed in quartz tubes with J. Young valves. Frozen solution EPR spectra were recorded on a JEOL continuous wave spectrometer JESFA200 equipped with an X-band Gunn diode oscillator bridge, a

3.5 Experimental

cylindrical mode cavity, and a helium cryostat. For all samples, a modulation frequency of 100 kHz and a time constant of 0.1 s were employed. All spectra were obtained on freshly prepared solutions of 1–10 mM compound in acetonitrile.

3.5.2 Computational Details

The program package ORCA 2.7 revision 0 was used for all calculations [18]. The geometry optimization calculations were performed by the spin-unrestricted DFT method with the BP86 [19–21] functional. The single point calculations and calculations of Mößbauer parameters were performed with the B3LYP functional [22–24]. The triple-ζ basis sets with one-set of polarization functions [25] (TZVP) were used for iron ions and the double-ζ basis sets with one-set of polarization functions [26] (SVP) were used for all other atoms. For calculation of Mößbauer parameters, the "core" CP(PPP) basis set for iron [27, 28] was used. This basis is based on the TurboMole DZ basis, developed by Ahlrichs et al. and obtained from the basis set library under ftp.chemie.uni-karlsruhe.de/pub/basen. Molecular orbitals were visualized via the program Molekel [29].

3.5.3 Synthetic Details

[**H$_3$TIMEN^{tol}**]**(BF$_4$)$_3$ (1(BF$_4$)$_3$).** A 100 mL flask was charged with *tris*(2-chloroethyl) amine (4.06 g, 19.86 mmol) and 1-(tolyl)imidazole (9.43 g, 59.59 mmol), the mixture was heated to 150 °C for 2 days. The resulting brown solid was dissolved in 20 mL of methanol and the hygroscopic chloride salt was converted to the corresponding stable tetrafluoroborate salt, [H$_3$TIMEN^{tol}](BF$_4$)$_3$, by addition of a solution of NaBF$_4$ (6.41 g, 60.00 mmol) in 50 mL of methanol. The tetrafluoroborate salt precipitated immediately, was collected by filtration, washed with methanol and diethyl ether, and dried *in vacuo*. (9.19 g, 11.02 mmol, yield 56%). Elemental analysis (%) for C$_{36}$H$_{44}$B$_3$F$_{12}$N$_7$O calcd. C 50.80, H 5.21, N 11.52; obsd: C 50.90, H 5.08, N 11.18. ^{1}H-NMR (270 MHz, DMSO-d_6, RT): δ [ppm] = 9.66 (s, 3 H), 8.16 (d, 3 H, ^3J = 1.5 Hz), 7.85 (d, 3 H, ^3J = 1.5 Hz), 7.54 (d, 6 H, ^3J = 8.6 Hz), 7.37 (d, 6 H, ^3J = 8.0 Hz), 4.36 (t, 6 H, ^3J = 6.5 Hz), 3.15 (s, 6 H, ^3J = 6.8 Hz), 2.38 (s, 9 H). ^{13}C{^{1}H}-NMR (100 MHz, DMSO-d_6, RT): δ (ppm) = 214.93 (3 C), 135.17 (3 C), 132.17 (3 C), 130.52 (6 C), 123.38 (3 C), 121.32 (6 C), 120.77 (3 C), 51.32 (3 C), 46.13 (3 C), 20.57 (3 C).

[**TIMEN^{tol}**] (**2**). A solution of potassium *tert*-butoxide (340 mg, 3.03 mmol) in 10 mL THF was added to a suspension of [H$_3$TIMEN^{tol}](BF$_4$)$_3$ (722 mg, 0.87 mmol) in 5 mL of THF and stirred for 3 h. The solution was then evaporated to dryness and the solid residue was dissolved in 15 mL of diethyl ether. The resulting solution was filtered through Celite and evaporated to dryness *in vacuo*. ^{1}H-NMR (270 MHz, RT, benzene-d_6): δ (ppm) = 7.74 (d, 6 H, ^3J = 8.5 Hz),

6.90 (d, 6 H, $^3J = 8.5$ Hz), 6.82 (d, 3 H, $^3J = 2.1$ Hz), 6.43 (d, 3 H, $^3J = 2.1$ Hz), 3.87 (t, 6 H, $^3J = 6.2$ Hz), 2.70 (t, 6 H, $^3J = 6.2$ Hz), 2.00 (s, 9 H). $^{13}C\{^1H\}$-NMR (100.5 MHz, benzene-d_6, RT): δ (ppm) = 214.61 (3 C), 140.71 (3 C), 135.28 (3 C), 129.88 (6 C), 121.22 (3 C), 120.91 (6 C), 116.55 (3 C), 56.41 (3 C), 49.73 (3 C), 20.71 (3 C).

[(TIMEN^{tol})Fe](BF$_4$)$_2$ (3). A: [H$_3$TIMEN^{tol}](BF$_4$)$_3$ (1,000 mg, 1.21 mmol) and sodium *tert*-butoxide (400 mg, 1.21 mmol) in 15 mL THF were stirred for 2 h, filtered and added to a suspension of FeCl$_2$ (135 mg, 1.21 mmol) in 5 mL pyridine. The reaction mixture was allowed to stir overnight, during which time an off-white precipitate formed. The precipitate was collected by filtration, washed with pyridine, diethyl ether and *n*-pentane, and dried *in vacuo*. **B:** A mixture of [H$_3$TIMEN^{tol}](BF$_4$)$_3$ (1,000 mg, 1.21 mmol), sodium *tert*-butoxide (400 mg), NaBPh$_4$ (410 mg, 1.21 mmol) and FeCl$_2$ (135 mg, 1.21 mmol) in THF was allowed to stir overnight, during which time an off-white precipitate formed. The precipitate was collected by filtration, washed with THF, diethyl ether and *n*-pentane, yielding the crude product. Recrystallization can be achieved by dissolving the crude product in acetonitrile and layering with diethyl ether at room temperature. Elemental analysis (%) for C$_{36}$H$_{39}$B$_2$F$_8$FeN$_7$ calcd. C 54.10, H 4.92, N 12.27; obsd: C 53.69, H 5.07, N 12.29. In the ^{1}H-NMR spectrum, eleven signals are expected for the complex, five are observed; the remaining signals are likely shifted and broadened into the baseline due to the paramagnetism of the compound. ^{1}H-NMR (270 MHz, RT, DMSO-d_6): δ (ppm) = -0.93, 10.62, 12.20, 14.68, 51.03. Mößbauer parameters: $\delta = 0.51(1)$ mm s^{-1}, $\Delta E_Q = 1.54(1)$ mm s^{-1}, $\Gamma_{FWHM} = 0.28(1)$ mm s^{-1}

[(TIMEN^{tol***})Fe] (4). A solution of [(TIMEN^{tol})Fe](BF$_4$)$_2$ (155 mg, 0.19 mmol) in THF was stirred over an excess of sodium amalgam (3,300 mg, Na Wt. 0.8%) overnight. The resulting red solution and precipitate were separated from the sodium amalgam with the help of a Pasteur pipette, the THF was removed and after addition of CHCl$_3$ the resulting solution was filtered through Celite. Removal of the solvent yields the crude product. Recrystallization can be achieved by dissolving the crude product in a mixture of CHCl$_3$ and benzene and cooling the filtered solution to -35 °C.

EPR (acetonitrile, 7.1 K, 8.9875 GHz, ModWidth = 1.0 mT, Power = 1.00 mW) $g_{\parallel} = 2.29$, $g_{\perp} = 1.94$. Mößbauer parameters: $\delta = 0.03(1)$ mm s^{-1}, $\Delta E_Q = 2.37(1)$ mm s^{-1}, $\Gamma_{FWHM} = 0.26(1)$ mm s^{-1}.

[H$_3$TIMEN^{3,5xyl}](PF$_6$)$_3$ (5(PF$_6$)$_3$). A 50 mL flask was charged with *tris*(2-chloroethyl)amine (8.49 g, 41.50 mmol) and 1-(3,5-xylyl)imidazole (22.00 g, 128.00 mmol), the mixture was heated to 150 °C for 2 days. The resulting brown solid was dissolved in 30 mL of methanol and the hygroscopic chloride salt was converted to the corresponding stable hexafluorophosphate salt, [H$_3$TIMEN^{3,5xyl}](PF$_6$)$_3$, by addition of a solution of NH$_4$PF$_6$ (22.00 g, 135.00 mmol) in 150 mL of methanol. The white hexafluorophosphate salt precipitated immediately, was collected by filtration, washed with methanol and diethyl ether, and the solid was dried *in vacuo*. (28.39 g, 27.00 mmol, yield 65%), ^{1}H-NMR (400 MHz, RT, DMSO-d_6): δ (ppm) = 9.72 (s, 3 H), 8.16 (s, 3 H), 7.88 (s, 3 H), 7.34 (s, 6 H), 7.25 (s, 3 H), 4.39 (t, 6 H, $^3J = 7.48$ Hz), 3.20 (t, 6 H, $^3J = 7.48$ Hz), 2.35

(s, 18 H). $^{13}C\{^1H\}$-NMR (100.5 MHz, RT, DMSO-d_6): δ (ppm) = 139.80 (6 C), 135.12 (3 C), 134.29 (3 C), 130.92 (3 C), 123.26 (3 C), 120.60 (3 C), 118.83 (6 C), 50.85 (3 C), 48.02 (3 C), 20.56 (6 C).

[**TIMEN^{3,5xyl}**] (**6**). A solution of potassium *tert*-butoxide (0.33 g, 2.95 mmol) in THF was added to a suspension of [H$_3$TIMEN^{3,5xyl}](PF$_6$)$_3$ (1.00 g, 0.95 mmol) in 5 mL of THF and stirred for 1 h. The solution was then evaporated to dryness and the solid residue was dissolved in 15 mL of diethyl ether. The resulting solution was filtered through Celite and the filtrate was evaporated to dryness *in vacuo*. ^{1}H-NMR (270 MHz, RT, benzene-d_6): δ (ppm) = 7.54 (s, 6 H), 6.88 (m, 3 H), 6.67 (m, 6 H), 6.63 (d, 3 H, ^3J = 1.32 Hz), 3.99 (t, 6 H ^3J = 6.24 Hz), 2.84 (t, 6 H, J = 6.24 Hz), 2.12 (s, 18 H). $^{13}C\{^1H\}$-NMR (100.5 MHz, RT, benzene-d_6): δ (ppm) = 212.70 (3 C), 142.46 (3 C), 138.55 (6 C), 128.33 (3 C) 121.07 (3 C), 118.95 (6 C), 116.48 (3 C), 56.15 (3 C), 49.38 (3 C), 21.02 (6 C).

[**(TIMEN^{3,5xyl})Fe(CH$_3$CN)](PF$_6$)$_2$** (**7**). [H$_3$TIMEN^{3,5xyl}](PF$_6$)$_3$ (1.00 g, 0.95 mmol) and potassium *tert*-butoxide (0.33 g, 2.95 mmol) in 15 mL THF were stirred for 2 h, filtered and added to a suspension of FeCl$_2$ (0.12 g, 0.95 mmol) in 5 mL pyridine. The reaction mixture was allowed to stir overnight, during which time an off-white precipitate formed. The precipitate was collected by filtration, washed with pyridine, diethyl ether and *n*-pentane, and dried *in vacuo*. Recrystallization of [(TIMEN^{3,5xyl})Fe](PF$_6$)$_2$ from a solvent mixture of acetonitrile and diethyl ether yielded the compound [(TIMEN^{3,5xyl})Fe(CH$_3$CN)](PF$_6$)$_2$ with a coordinated acetonitrile molecule (0.64 g, 0.66 mmol, yield 70%). Elemental analysis (%) for C$_{41}$H$_{48}$F$_{12}$FeN$_8$P$_2$ calcd. C 49.31, H 4.84, N 11.22; obsd: C 49.46, H 4.85, N 11.42. In the ^{1}H-NMR, twelve signals are expected for the complex, five are observed, the remaining signals are likely shifted and broadened into the baseline due to the paramagnetism of the compound. ^{1}H-NMR (270 MHz, RT, DMSO-d_6): δ (ppm) = 23.99, 20.82, 12.88, 10.29, 4.50. Mößbauer parameters for [(TIMEN^{3,5xyl})Fe](PF$_6$)$_2$ δ = 0.52(1) mm s^{-1}, ΔE_Q = 1.12(1) mm s^{-1}, Γ_{FWHM} = 0.50(1) mm s^{-1}. Mößbauer parameters for [(TIMEN^{3,5xyl})Fe(CH$_3$CN)](PF$_6$)$_2$ δ = 0.73(1) mm s^{-1}, ΔE_Q = 1.08(1) mm s^{-1}, Γ_{FWHM} = 0.36(1) mm s^{-1}.

[**(TIMEN^{3,5xyl**})Fe**] (**8**). A solution of [(TIMEN^{3,5xyl})Fe](PF$_6$)$_2$ (0.30 g, 0.31 mmol) in THF was stirred overnight over an excess of sodium amalgam (5.60 g, Na Wt. 0.8%). The solution was filtered through Celite, concentrated *in vacuo* and cooled to -35 °C to yield the product as dark red crystals. Mößbauer parameters: δ = 0.07(1) mm s^{-1}, ΔE_Q = 2.37(2) mm s^{-1}, Γ_{FWHM} = 0.47(3) mm s^{-1}.

[**(TIMEN^{3,5xyl**})Fe](PF$_6$)** (**9**). **A**: [(TIMEN^{3,5xyl**})Fe] (200 mg, 0.30 mmol) was dissolved in 10 mL CHCl$_3$ to form a green solution. Excess of NaPF$_6$ was added and diethyl ether was then diffused into the filtered solution at room temperature to give green crystals overnight. **B**: Addition of silver triflate (77 mg, 0.30 mmol) to a suspension of [(TIMEN^{3,5xyl**})Fe] (200 mg, 0.30 mmol) yielded a green solution and a dark grey precipitate of elemental silver. The solution was filtered through Celite and the solvent was removed at reduced pressure to yield the green product.

EPR (acetonitrile, 86 K, 8.9817 GHz, Mod Width = 1.0 mT, Power = 1.00 mW) g_1 = 2.42, g_2 = 2.20, g_3 = 1.94.

78 3 TIMEN$^{tol/3,5xyl}$: Unexpected Reactivity Resulting From Modifications

Table 3.5 Crystallographic data, data collection, and refinement details of [H$_3$TIMEN^{tol}] (CF$_3$SO$_3$)$_3$ (**1**(CF$_3$SO$_3$)$_3$), [(TIMEN^{tol})Fe](BF$_4$)$_2$ · 0.8 Et$_2$O (**3** · 0.8 Et$_2$O), and [(TIMEN^{tol***})Fe] · 2 THF (**4** · 2 THF)

	1(CF$_3$SO$_3$)$_3$ [H$_3$TIMEN^{tol}](CF$_3$SO$_3$)$_3$	**3** · 0.8 Et$_2$O [(TIMEN^{tol})Fe](BF$_4$)$_2$ · 0.8 Et$_2$O	**4** · 2 THF [(TIMEN^{tol***})Fe] · 2 THF
Empirical formula	C$_{39}$H$_{42}$F$_9$N$_7$O$_9$S$_3$	C$_{39.2}$H$_{47}$B$_2$F$_8$FeN$_7$O$_{0.8}$	C$_{44}$H$_{52}$FeN$_7$O$_2$
Mol. weight	1,019.98	858.51	766.78
Crystal size (mm^3)	0.37 × 0.11 × 0.10	0.24 × 0.18 × 0.07	0.20 × 0.10 × 0.08
Temperature (K)	100	100	150
Crystal system	Triclinic	Monoclinic	Triclinic
Space group	$P\bar{1}$(no. 2)	$P2_1/c$ (no. 14)	$P\bar{1}$(no. 2)
a (Å)	10.2584(7)	14.761(2)	12.355(1)
b (Å)	12.483(2)	13.986(2)	13.3642(8)
c (Å)	19.679(2)	20.053(3)	13.7659(7)
α (°)	73.265(10)	90	106.685(4)
β (°)	84.479(8)	100.39(2)	90.960(5)
γ (°)	67.943(2)	90	116.281(5)
V (Å^3)	2,236.4(5)	4,071.9(9)	1,924.7(2)
Z	2	4	2
ρ_{calc} (g cm^{-3})	1.515	1.400	1.323
μ (mm^{-1})	0.265	0.448	0.440
F (000)	1,052	1,782	814
T_{min}; T_{max}	0.897; 0.970	0.846; 0.970	0.842; 0.965
2θ interval (°)	$6.7 \leq 2\theta \leq 54.2$	$6.6 \leq 2\theta \leq 54.2$	$6.2 \leq 2\theta \leq 54.2$
Collected reflections	66,332	85,162	53,280
Independent reflections; R_{int}	9,846; 0.0441	8,955; 0.0667	8,478; 0.0371
Observed reflections [$I \geq 2\sigma(I)$]	7,902	7,000	7,410
No. refined parameters	671	637	536
wR_2 (all data)	0.0900	0.1159	0.1223
R_1 [$I \geq 2\sigma(I)$]	0.0366	0.0449	0.0428
GooF on F^2	1.023	1.044	1.052
$\Delta\rho_{max/min}$	0.527/−0.358	0.495/−0.458	0.917/−0.587

Table 3.6 Crystallographic data, data collection, and refinement details of $[H_3TIMEN^{3,5xyl}](Cl)_3 \cdot$ solv ($5(Cl)_3 \cdot$ solv), $[(TIMEN^{3,5xyl})Fe(CH_3CN)](PF_6)_2 \cdot 2$ CH_3CN ($7\ 2\ CH_3CN$), $[(TIMEN^{3,5xyl**})Fe]$ (**8**), and $[(TIMEN^{3,5xyl**})Fe](PF_6) \cdot 0.6$ THF $\cdot 0.4$ $CHCl_3$ (**9** $\cdot 0.6$ THF $\cdot 0.4$ $CHCl_3$)

	$5(Cl)_3 \cdot$ solv $[H_3TIMEN^{3,5xyl}]$ $(Cl)_3 \cdot$ solv	$7 \cdot 2$ CH_3CN $[(TIMEN^{3,5xyl})Fe$-$(CH_3CN)](PF_6)_2 \cdot 2$ CH_3CN	**8** $[(TIMEN^{3,5xyl**})Fe]$	**9** $\cdot 0.6$ THF $\cdot 0.4$ $CHCl_3$ $[(TIMEN^{3,5xyl**})Fe](PF_6) \cdot 0.6$ THF $\cdot 0.4$ $CHCl_3$
Empirical formula	$C_{39.9375}H_{53.1875}Cl_{5.8125}N_7O_{2.125}$	$C_{45}H_{54}F_{12}FeN_{10}P_2$	$C_{39}H_{43}FeN_7$	$C_{41.8}H_{48.2}Cl_{1.2}F_6Fe$ $N_7O_{0.6}P$
Mol. weight	871.39	1,080.77	665.65	901.63
Crystal size (mm³)	$0.28 \times 0.16 \times 0.04$	$0.50 \times 0.23 \times 0.18$	$0.24 \times 0.08 \times 0.08$	$0.20 \times 0.06 \times 0.02$
Temperature (K)	100	150	150	100
Crystal system	Triclinic	Triclinic	Triclinic	Monoclinic
Space group	$P\bar{1}$(no. 2)	$P\bar{1}$(no. 2)	$P\bar{1}$(no. 2)	$P2_1/c$ (no. 14)
a (Å)	17.660(4)	11.488(2)	11.5632(7)	14.001(2)
b (Å)	21.259(5)	14.968(2)	11.9611(7)	13.587(2)
c (Å)	26.505(6)	15.908(2)	12.3184(9)	23.134(4)
α (°)	106.629(4)	74.453(7)	81.472(5)	90
β (°)	104.166(4)	88.036(8)	71.432(5)	102.443(3)
γ (°)	94.311(4)	70.889(7)	82.705(5)	90
V (Å³)	9,131(4)	2,485.7(4)	1,591.4(2)	4,297(2)
Z	8	2	2	4
ρ_{calc} (g cm⁻³)	1.341	1.444	1.389	1.394
μ(mm⁻¹)	0.406	0.457	0.516	0.530
F (000)	3,661	1,116	704	1,873
T_{min}; T_{max}	0.665; 0.746	0.739; 0.921	0.818; 0.960	0.506; 0.421
2θ interval (°)	$6.0 \leq 2\theta \leq 51.4$	$6.5 \leq 2\theta \leq 54.2$	$6.7 \leq 2\theta \leq 54.2$	$5.1 \leq 2\theta \leq 54.2$

(continued)

Table 3.6 (continued)

	5(Cl)$_3$ · solv [H$_3$TIMEN^{3,5xyl}](Cl)$_3$ · solv	7 · 2 CH$_3$CN [(TIMEN^{3,5xyl})Fe-(CH$_3$CN)](PF$_6$)$_2$ · 2 CH$_3$CN	8 [(TIMEN^{3,5xyl**})Fe]	9 · 0.6 THF · 0.4 CHCl$_3$ [(TIMEN^{3,5xyl**})Fe](PF$_6$) · 0.6 THF · 0.4 CHCl$_3$
Collected reflections	134,232	70,427	45,758	33,178
Independent reflections; R_{int}	34,579; 0.0905	10,940; 0.0725	7,010; 0.0711	9,448; 0.0697
Observed reflections [$I \geq 2\sigma(I)$]	17,923	8,332	5,529	6,073
No. refined parameters	2,248	704	430	612
wR_2 (all data)	0.2667	0.1269	0.0901	0.1285
R_1 [$I \geq 2\sigma(I)$]	0.0892	0.0467	0.0397	0.0506
GooF on F^2	1.341	1.039	1.025	1.010
$\Delta\rho_{max/min}$	1.811/−1.203	0.696/−0.577	0.368/−0.369	0.506/−0.421

3.5 Experimental

3.5.4 X-ray Crystal Structure Determination Details

Colorless plates of $[H_3TIMEN^{tol}](CF_3SO_3)_3$ ($1(CF_3SO_3)_3$) were grown by layering an n-pentane solution of the corresponding imidazolium salt with diethyl ether. Colorless plates of $[(TIMEN^{tol})Fe](BF_4)_2 \cdot 0.8$ Et$_2$O ($3 \cdot 0.8$ Et$_2$O) were obtained from a solution of the complex in a mixture of acetonitrile and diethyl ether. Dark red needles of $[(TIMEN^{tol***})Fe] \cdot 2$ THF ($4 \cdot 2$ THF) were grown from a THF solution of the complex at -35 °C. Colorless plates of $[H_3TIMEN^{3,5xyl}]$ (Cl)$_3 \cdot$ solv [5(Cl)$_3 \cdot$ solv (solv $= 0.9735$ CHCl$_3 \cdot 2.125$ H$_2$O)] precipitated at room temperature from a CHCl$_3$ solution of the hygroscopic imidazolium salt layered with diethyl ether. Brown irregular shaped crystals of $[(TIMEN^{3,5xyl})$ Fe(CH$_3$CN)](PF$_6$)$_2 \cdot 2$ CH$_3$CN ($7 \cdot 2$ CH$_3$CN) were obtained by slow diethyl ether diffusion into a layered acetonitrile solution of the complex. Dark red plates of $[(TIMEN^{3,5xyl**})Fe]$ (8) were grown from a THF/n-pentane mixture. Brown plates of $[(TIMEN^{3,5xyl**})Fe](PF_6) \cdot 0.6$ THF $\cdot 0.4$ CHCl$_3$ ($9 \cdot 0.6$ THF $\cdot 0.4$ CHCl$_3$) were grown from a mixture of chloroform, dichloromethane and THF. Suitable single crystals of the compounds were embedded in protective perfluoropolyalkylether oil and quickly transferred to the cold nitrogen gas stream of the diffractometer. Intensity data were collected either on a Bruker-Nonius KappaCCD diffractometer [$1(CF_3SO_3)_3$, $3 \cdot 0.8$ Et$_2$O, $4 \cdot 2$ THF, $7 \cdot 2$ CH$_3$CN and 8] using graphite monochromatized MoK_λ radiation ($\lambda = 0.71073$ Å) or on a Bruker Kappa APEX2 Duo diffractometer (5(Cl)$_3 \cdot$ solv and $9 \cdot 0.6$ THF $\cdot 0.4$ CHCl$_3$) equipped with an $I\lambda S$ microsource and QUAZAR focusing Montel optics using MoK_λ radiation ($\lambda = 0.71073$ Å). Lorentz and polarization effects were taken into account during data reduction, semiempirical absorption corrections were performed on the basis of multiple scans using $SADABS$ [30]. All structures were solved by direct methods and refined by full-matrix least-squares procedures on F^2 (with the exception of $5 \cdot$ solv, which was refined in a block matrix) using $SHELXTL$ NT 6.12 [31].

In $[H_3TIMEN^{tol}](CF_3SO_3)_3$ ($1(CF_3SO_3)_3$), one of the imidazolium rings is disordered. Two preferred orientations were refined resulting in occupancies of 86.9(3)% for atoms N2, N3, C1–C5 and 13.1(3)% for N2A, N3A, C1A–C5A. SIMU and ISOR restraints were applied. In $[(TIMEN^{tol})Fe](BF_4)_2 \cdot 0.8$ Et$_2$O ($3 \cdot 0.8$ Et$_2$O) both BF$_4$ anions are subjected to rotational disorder around one B–F bond. Two preferred orientations were refined each resulting in occupancies of 57(2) and 43(2)% for atoms F12–F14 and F12A–F14A, respectively and of 52(5) and 48(5)% for atoms F22–F24 and F22A–F24A, respectively. $[(TIMEN^{tol***})$ Fe] $\cdot 2$ THF ($4 \cdot 2$ THF) crystallizes with two molecules of THF per formula unit one of which is disordered showing two alternative orientations of 46.4(7) and 53.6(7)% occupancy for O200–C204 and O210–C214, respectively. SIMU, ISOR and SAME restraints were applied in the refinement of the disorder. The asymmetric unit of $[H_3TIMEN^{3,5xyl}]$(Cl)$_3 \cdot$ solv (5(Cl)$_3 \cdot$ solv) contains a total of four independent molecules of the imidazolium salt, 3.75 molecules of CHCl$_3$ and 8.5 molecules of solvent water. In one of the imidazolium cations one of the three ligand arms is disordered. Two alternative positions were refined resulting in site occupancies of 81.4(2) and 18.6(2)% for the atoms N306, N307, C311–C315, C332–C339 and N406, N407, C411–C415, C432–C439, respectively. This disorder also affects some of the $CHCl_3$ solvate molecules while other solvate molecules are occupied by approximately 50 or 75%. SIMU, ISOR, DFIX, FLAT, and SAME restraints were applied in the refinement of $5(Cl)_3 \cdot$ solv. In [(TIMEN^{3,5xyl}) Fe(CH$_3$CN)](PF$_6$)$_2 \cdot 2$ CH$_3$CN (7 $\cdot 2$ CH$_3$CN), one of the PF$_6$ anions is disordered. Two alternative orientations were refined resulting in occupancies of 61.1(6) and 38.9(6)% for atoms P2–F26 and P2A–F26A, respectively. In [(TIMEN^{3,5xyl**})Fe](PF$_6$) $\cdot$ 0.6 THF $\cdot$ 0.4 CHCl$_3$ (9 $\cdot$ 0.6 THF $\cdot$ 0.4 CHCl$_3$), one of the PF$_6$ anions is subjected to rotational disorder in the equatorial plane. Two alternative orientations were refined resulting in occupancies of 59.8(6) and 40.2(6)% for atoms F13–F16 and F13A–F16A, respectively. Compound 9 $\cdot$ 0.6 THF $\cdot$ 0.4 CHCl$_3$ crystallizes with a mixture of solvent molecules (0.6 CHCl$_3$ and 0.4 THF) which share a common crystallographic site. SIMU, ISOR, and SAME restraints were applied in the refinement.

In the computations of the crystal structure determinations of compounds 1–9, all non-hydrogen atoms were refined with anisotropic displacement parameters. All hydrogen atoms were placed in positions of optimized geometry, their isotropic displacement parameters were tied to those of the corresponding carrier atoms by a factor of either 1.2 or 1.5 (Tables 3.5 and 3.6).

Acknowledgments Text, schemes, and figures of this chapter, in part, are reprints of the materials published in the following paper Vogel et al. [32]. The dissertation author was the primary researcher and author. The co-authors listed in the publication also participated in the research. The permission to reproduce the paper was granted by Elsevier. Copyright 2010, Elsevier.

References

1. W.B. Tolman (ed.), *Activation of Small Molecules: Organometallic and Bioinorganic Perspectives* (Wiley-VCH, Weinheim, 2006)
2. M. Alcarazo, T. Stork, A. Anoop, W. Thiel, A. Fürstner, Angew. Chem. Int. Ed. **49**, 2542 (2010)
3. X. Hu, I. Castro-Rodriguez, K. Olsen, K. Meyer, Organometallics **23**, 755 (2004)
4. F.E. Hahn, M.C. Jahnke, Angew. Chem. Int. Ed. **47**, 3122 (2008)
5. X. Hu, K. Meyer, J. Organomet. Chem. **690**, 5474 (2005)
6. O. Einsle, F.A. Tezcan, S.L.A. Andrade, B. Schmid, M. Yoshida, J.B. Howard, D.C. Rees, Science **297**, 1696 (2002)
7. M. Aliaga-Alcalde, S.D. George, B. Mienert, E. Bill, K. Wieghardt, F. Neese, Angew. Chem. Int. Ed. **44**, 2908 (2005)
8. C. Vogel, F.W. Heinemann, J. Sutter, C. Anthon, K. Meyer, Angew. Chem. Int. Ed. **47**, 2681 (2008)
9. J.J. Scepaniak, M.D. Fulton, R.P. Bontchev, E.N. Duesler, M.L. Kirk, J.M. Smith, J. Am. Chem. Soc. **130**, 10515 (2008)
10. J.J. Scepaniak, J. Young, R. Bontchev, J.M. Smith, Angew. Chem. Int. Ed. **48**, 3158 (2009)

References

11. C.S. Vogel, F.W. Heinemann, M.M. Khusniyarov, K. Meyer, Inorg. Chim. Acta. **364**, 226 (2010)
12. X. Hu, K. Meyer, J. Am. Chem. Soc. **126**, 16322 (2004)
13. A.J. Arduengo III, F.P. Gentry Jr, P.K. Taverkere, H.E. Simmons III. (E. I. Du Pont de Nemours & Co., USA) US 6177575, **7** (2001)
14. K. Ward Jr, J. Am. Chem. Soc. **57**, 914 (1935)
15. X. Hu, I. Castro-Rodriguez, K. Meyer, J. Am. Chem. Soc. **126**, 13464 (2004)
16. M. Brookhart, M.L.H. Green, G. Parkin, Proc. Natl. Acad. Sci. **104**, 6908 (2007)
17. P.B. Hitchcock, M.F. Lappert, P. Terreros, J. Organomet. Chem. **239**, C26 (1982)
18. F. Neese, *ORCA—an Ab Initio, Density Functional and Semiempirical SCF-MO Package, version 2.7 revision 0; Institut für Physikalische und Theoretische Chemie* (Universität Bonn, Germany, 2009)
19. A.D. Becke, Phys. Rev. A **38**, 3098 (1988)
20. J.P. Perdew, Phys. Rev. B **33**, 8822 (1986)
21. J.P. Perdew, Phys. Rev. B **34**, 7406 (1986)
22. A.D. Becke, J. Chem. Phys. **98**, 5648 (1993)
23. C.T. Lee, W.T. Yang, R.G. Parr, Phys. Rev. B **37**, 785 (1988)
24. P.J. Stephens, F.J. Devlin, C.F. Chabalowski, M.J. Frisch, J. Phys. Chem. 98, 11623 (1994)
25. D. Bourissou, O. Guerret, F.P. Gabbaie, G. Bertrand, Chem. Rev. **100**, 39 (2000)
26. A. Schäfer, H. Horn, R. Ahlrichs, J. Chem. Phys. **97**, 2571 (1992)
27. F. Neese, Inorg. Chim. Acta **337**, 181 (2002)
28. S. Sinnecker, L.D. Slep, E. Bill, F. Neese, Inorg. Chem. **44**, 2245 (2005)
29. S. Portmann, Molekel, version 4.3.win32, CSCS/UNI Geneva, Switzerland (2002)
30. SADABS 2.06, Bruker AXS, Inc., Madison WI., U.S.A (2002)
31. SHELXTL NT 6.12, Bruker AXS, Inc., Madison WI., U.S.A (2002)
32. C.S. Vogel, F.W. Heinemann, M.M. Khusniyarov, K. Meyer, Unexpected reactivity resulting from modifications of the ligand periphery: Synthesis, structure, and spectroscopic properties of iron complexes of new tripodal N-heterocyclic carbene (NHC) ligands, Inorganica Chimica Acta 364, 226–237 (2010) (invited article to special issue in honor of Prof. Dr. A.L. Rheingold)

Chapter 4
TIMEN^{3,5CF3}: Hybrid System Capable of Cyclometallation and Nitride Insertion

4.1 Introduction

In an effort to develop coordinatively unsaturated, electron-rich metal complexes, that have proven to be powerful species for small molecule activation [1], a new derivative of the tripodal NHC ligand family TIMEN^R (*tris*-[2-(3-aryl-imidazol-2-ylidene)ethyl]amine, with R = aryl) was synthesized. Due to their σ- donor and π-accepting properties [2, 3], N-heterocyclic carbene (NHC) ligands are exceedingly suitable for the synthesis of a variety of low to high-valent metal complexes; thus, perfectly suitable to act as supporting ligands for small molecule activation at reactive coordination complexes [4]. The steric demand of the NHC ligands is highly adjustable, and often, the ligands can be conveniently synthesized. Slight changes in the steric demand of the bulky ligand periphery can be used to tune the reactivity of the resulting transition metal complexes. The anchoring unit of polydentate NHC systems provides the chelators with additional electronic and structural flexibility. It could be shown that sterically protecting tripodal ligands of TIMEN^R (with R = aryl = 2,6-xylyl (xyl), mesityl (mes)) create a trigonal platform for metal ions that enables the coordination of an axial ligand in a protective cylindrical cavity, such as the nitrido ligand. The first X-ray crystal structure determination of a discrete iron nitride complex, [(TIMEN^R) Fe(N)]$^+$, stabilized by the N-anchored *tris*(carbene) ligand TIMEN^R was accomplished [5]. Shortly thereafter, the second X-ray crystallographic characterization of an iron nitride complex with a closely related borate-anchored *tris*(carbene) ligand was reported by Smith et al. i.e. [(PhB^{tBu}Im)Fe(N)] [6] showing the great potential of carbene based ligand systems, they enable both the generation and stabilization of reactive iron nitride complexes. In order to synthesize N-anchored Fe$\equiv$N complexes for reactivity studies, we considered different options. One possible approach was the oxidation of the Fe (IV) nitride complex, [(TIMEN^{mes})Fe(N)] BPh$_4$, to yield an Fe (V) species. But this attempt led unexpectedly to reduced *bis*-carbene imine species, [(TIMEN^{mes} = N^H)Fe(NCMe)] BPh$_4$ (see Chap. 2), that formed via

C. S. Vogel, *High- and Low-Valent* tris-*N-Heterocyclic Carbene Iron Complexes*,
Springer Theses, DOI: 10.1007/978-3-642-27254-7_4,
© Springer-Verlag Berlin Heidelberg 2012

86   4 TIMEN^{3,5CF3}: Hybrid System Capable of Cyclometallation and Nitride Insertion

insertion of the terminal nitrido ligand into one of the iron–carbene bonds of the TIMEN^{mes} ligand. This reaction pathway appears to be reminiscent of the insertion chemistry observed for the Co(III) imide complex, [(TIMEN^R)Co(NR)]BPh4 [7]. Even in the presence of substrates, these examples of observed intramolecular insertion chemistry showed that the methyl groups in *ortho* position of the mesityl rings of the NHC ligand play an important role in the reaction behavior of the corresponding iron complex. Due to a strong steric protection of the substituents on these positions, side-access of potential substrates to the axial $M \equiv L$ functional group is efficiently suppressed, thereby preventing further reactivity of the $Fe \equiv N$ unit in [(TIMEN^{mes})Fe(N)]BPh$_4$ and its oxidized Fe (V) species. Another approach was to modify our NHC systems by varying the steric demand of the aryl substituents. Two new derivatives of the TIMEN^R family (TIMEN^R, with R = aryl = tolyl (tol), 3,5-xylyl (3,5xyl)) have been discussed in Chap. 3 [8]. Investigations of these derivatized ligands showed that relatively small changes in the periphery of the ligand have drastic effects on the reactivity of the resulting complexes. Attempts to reduce the Fe(II) precursors, [(TIMEN$^{tol/3,5xyl}$) Fe], over sodium amalgam did not yield the expected Fe(I) complexes. Instead, cyclometallation of the reactive ortho C–H groups of the aryl carbene moieties occurred leading to pseudo-octahedrally coordinated, *tris*-metallated (in the case of the tolyl system) or *bis*-metallated (for the 3,5-xylyl system) Fe(II) and Fe(III) complexes.

The new TIMEN^{3,5CF3} system can be considered as hybrid between the two different kinds of ligand systems, namely TIMEN^{mes} and TIMEN$^{tol/3,5xyl}$, because it is capable of both nitride insertion and cyclometallation. Actually, these two reaction occur in one-pot synthesis initiated simply by photolysis of the Fe(II) azide precursor. The synthesis and characterization of a new N-anchored NHC ligand of the TIMEN^R system, that is to say the fluorinated derivative TIMEN^{3,5CF3}, as well as the resulting iron coordination compounds, [(TIMEN^{3,5CF3})Fe(Cl)] (Cl) (**3**), [(TIMEN^{3,5CF3})Fe(N$_3$)] (BPh$_4$) (**4**), and [(TIMEN^{3,5CF3}*)(*NH)Fe(NCMe)](BPh$_4$) (**5**) (see Chart. 4.1) are presented.

4.2 Results and Discussion

The general synthetic route of the tripodal ligand system TIMEN^R, as well as the generation of the carbenes and their coordination to iron were reported in the previous chapters [8]. The protocol for the TIMEN^{3,5CF3} is similar. Deprotonation of the imidazolium salt with a strong base, like potassium *tert*-butoxide, yields the N-heterocyclic carbene. Reaction of the free carbene with ferrous chloride in THF delivers the five-coordinated Fe(II) complex, [(TIMEN^{3,5CF3})Fe(Cl)](Cl) (**3**) as analytically pure, yellow-greenish powder in 85% yield.

The ^{1}H-NMR of **3** in solution is consistent with a paramagnetic high-spin iron(II) complex (Fig. 4.7). Complex **3** exhibits a Mößbauer isomer shift, δ of 0.75(1) mm s^{-1}, which is characteristic but slightly more positve than other Fe(II) high-spin complexes of this system ($\delta \approx$ 0.65–0.73 mm s^{-1}, see Chaps. 2, 3) [5].

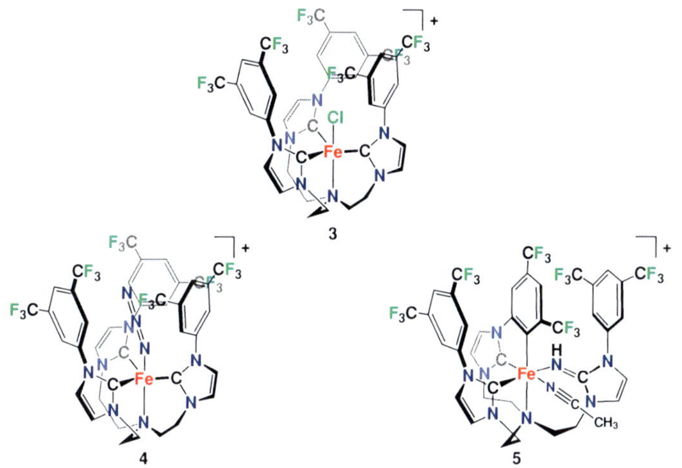

4.2 Results and Discussion

Chart. 4.1 Overview of the iron complexes derived from the tripodal TIMEN^{3,5CF3} ligand

The quadrupole splitting, ΔE_Q of 3.04(1) mm s^{-1} is relative large compared to other Fe(II) high-spin complexes bearing the TIMEN^R ligand ($\Delta E_Q \approx$ 1.08–2.34 mm s^{-1}, see Chaps. 2, 3). The observed quadrupole splitting, ΔE_Q, is directly correlated with the molecule symmetry as well as with the configuration of the valence electrons influenced by the ligand field effect in transition metal complexes. ΔE_Q derives from the electric quadrupole interaction between the electric quadrupole moment of a nucleus and the inhomogeneous electric field at the position of this nucleus. In brief, ΔE_Q provides information on the molecule symmetry, the oxidation state, and bond characteristics [9]. One explanation for the relative large quadrupole splitting, ΔE_Q, together with the slightly larger isomer shift, δ, could be the more pronounced inhomogeneity of the electric field around the nucleus derived from disparate ligands, with regard to the ligand field strength, coordinated to iron. This is also reflected in the deviations of the bond distances around the iron center when comparing the iron(II) chloro complexes of the two different systems TIMEN^{3,5CF3} and TIMEN^{mes}. See Table 4.1 for the principle bond distances in the complex core structure. The differences in bond length, reflecting the bond strength, are larger in the [(TIMEN^{3,5CF3})Fe(Cl)](Cl) than in the [(TIMEN^{mes})Fe(Cl)]Cl case. It can be shown that, via introduction of electron withdrawing groups ($-CF_3$) with a negative inductive ($-$ I-effect) and mesomeric effect ($-$ M-effect) instead of electron donating groups ($-CH_3$) with a positive inductive ($+$ I-effect) and mesomeric effect ($+$ M-effect) at the phenyl rests, the electronic structure of the iron complexes can be influenced significantly (Figs. 4.1 and 4.3).

The molecular structure of the cation [(TIMEN^{3,5CF3})Fe(Cl)]$^+$ in light green crystals of [((TIMEN^{3,5CF3})Fe(Cl)](Cl) · KCl · 2 Et$_2$O (**3** · KCl · 2 Et$_2$O), obtained by diffusion method using the solvent mixture diethyl ether and dichloromethane, exhibits a four-coordinate iron center with three equivalent iron carbene distances

Table 4.1 Principle bond distances [Å] with e.s.d.'s in parentheses for [(TIMEN^{3,5CF3})Fe(Cl)](Cl) · KCl · 2 Et$_2$O (3 · KCl · 2 Et$_2$O), [(TIMEN^{3,5CF3}) Fe(N$_3$)] (BPh$_4$) · Et$_2$O (4 · Et$_2$O), [(TIMEN^{mes}) Fe(Cl)]Cl · 3 MeCN

	[(TIMEN^{3,5CF3})Fe(Cl)]$^+$	[(TIMEN^{3,5CF3})Fe(N$_3$)]$^+$	[(TIMEN^{mes})Fe(Cl)]$^+$
Fe1–N1	3.084(4)	3.082(5)	3.008(3)
Fe1–Lig$_{axial}$[a]	2.310(2)	1.966(4)	2.2768(8)
Fe1–C3	2.115(5)	2.116(4)	2.128(3)
Fe1–C8	2.115(5)	2.109(4)	2.138(3)
Fe1–C13	2.112(5)	2.122(4)	2.138(2)

[a] Lig$_{axial}$ corresponds to the respective axial Ligand, for compound **3** : Cl1 (chloro); for compound **4** : N8 (azide); for [(TIMEN^{mes})Fe(Cl)]Cl · 3 MeCN: Cl1

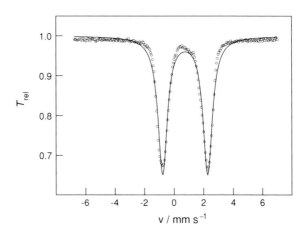

Fig. 4.1 Zero-field Mößbauer spectrum of [(TIMEN^{3,5CF3})Fe(Cl)](Cl) (**3**) (recorded in solid state at 77 K; $\delta = 0.75(1)$ mm s^{-1}, $\Delta E_Q = 3.04(1)$ mm s^{-1}, $\Gamma_{FWHM} = 0.75(1)$ mm s^{-1})

(average Fe–C$_{carbene}$ distance 2.114 Å). While the axial chloro ligand binds with a bond distance of 2.310(2) Å, the Fe–N$_{anchor}$ distance of 3.084(4) Å, indicates that the N-anchor is non-coordinating. The metal center is located 0.510(3) Å above the trigonal plane of the three NHC carbene carbon atoms (Fig. 4.2).

The corresponding Fe(II) azide complex is readily accessible via salt metathesis. Stirring the Fe(II) chloro precursor complex (**3**) over an excess of sodium azide in acetonitrile yields [(TIMEN^{3,5CF3})Fe(N$_3$)](N$_3$) (**4**).

^{1}H-NMR spectrum of **4** in solution is consistent with a paramagnetic high-spin iron(II) complex and very similar to **3** (Fig. 4.7). Also the Mößbauer parameters of the azide compound are very similar to those of the chloro precursor **3**. Complex **4** exhibits a Mößbauer isomer shift, δ of 0.73(1) mm s^{-1} and a quadrupole splitting, ΔE_Q of 3.04(1) mm s^{-1}. Comparing the complex core structure of **4** with **3** (Table 4.1) one can notice that Fe-C$_{carbene}$ distances are in the same range as well (Fig. 4.3).

The molecular structure of the cation [(TIMEN^{3,5CF3})Fe(N$_3$)]$^+$ in light orange crystals of [(TIMEN^{3,5CF3})Fe(N$_3$)](BPh$_4$) · Et$_2$O (**4** · Et$_2$O), obtained from a solvent mixture of Et$_2$O and MeCN, shows that upon exchange of the axial ligand the core complex barely changes (Fig. 4.4).

4.2 Results and Discussion

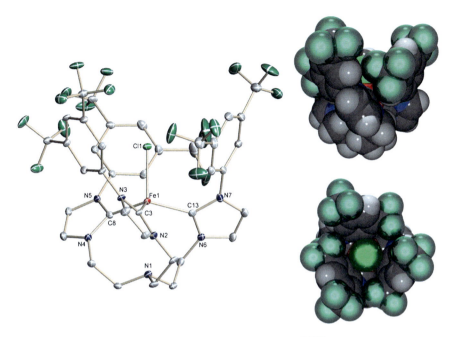

Fig. 4.2 Molecular structure of the complex cation of [((TIMEN^{3,5CF3})Fe(Cl)](Cl) (**3**) in crystals of **3** · KCl · 2 Et$_2$O (30 % probability ellipsoids). Hydrogen atoms, BPh$_4^-$ anion, and co-crystallized solvents are omitted for clarity. Space-filling representations of the cation **3** side view (*right, top*) and top view (*right, bottom*). Selected bond distances (Å): Fe1–N1 3.084(4), Fe1–Cl1 2.310(2), Fe1–C3 2.115(5), Fe1–C8 2.115(5), Fe1–C13 2.112(5), $d_{Fe\ oop}$ 0.510(3)

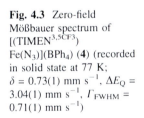

Fig. 4.3 Zero-field Mößbauer spectrum of [(TIMEN^{3,5CF3})Fe(N$_3$)](BPh$_4$) (**4**) (recorded in solid state at 77 K; δ = 0.73(1) mm s^{-1}, ΔE_Q = 3.04(1) mm s^{-1}, Γ_{FWHM} = 0.71(1) mm s^{-1})

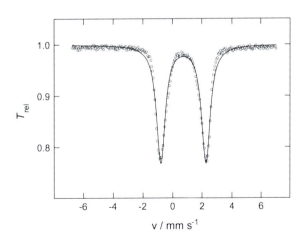

The iron center is again four-coordinated with three equivalent iron carbene distances (average Fe–C$_{carbene}$ distance 2.116 Å, Table 4.1). The axial azide ligand binds with a bond distance of 1.966(4) Å. The Fe–N$_{anchor}$ distance changes only slightly from 3.084(4) Å in the chloro species to 3.082(5) Å in the azide complex.

90   4 TIMEN^{3,5CF3}: Hybrid System Capable of Cyclometallation and Nitride Insertion

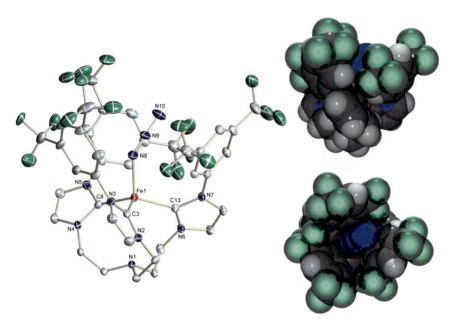

Fig. 4.4 Molecular structure of the complex cation of [(TIMEN^{3,5CF3})Fe(N$_3$)](BPh$_4$) (**4**) in crystals of **4** · Et$_2$O (30% probability ellipsoids). Hydrogen atoms, BPh$_4^-$ anion, and co-crystallized solvents are omitted for clarity. Space-filling representations of the cation **4** side view (*right, top*) and top view (*right, bottom*). Selected bond length (Å): Fe1–N8 1.966(4), Fe1–C3 2.116(4), Fe1–C8 2.109(4), Fe1–C13 2.122(4), N8–N9 1.074(16), N9–N10 1.171(17), $d_{Fe\ oop}$ 0.515(3)

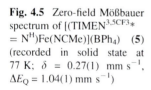

Fig. 4.5 Zero-field Mößbauer spectrum of [(TIMEN^{3,5CF3}$*$ = N^H)Fe(NCMe)](BPh$_4$) (**5**) (recorded in solid state at 77 K; δ = 0.27(1) mm s^{-1}, ΔE_Q = 1.04(1) mm s^{-1})

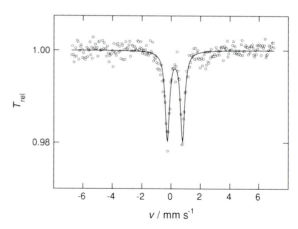

Notable is that the TIMEN^{3,5CF3} system, compared to the TIMEN$^{2,6xyl/mes}$ sytems [5], allows the azide ligand to coordinate in the usually observed bent fashion. The steric pressure exerted on the iron azide coordination by the aryl substituents is reduced in the case of the fluorinated species. Most probably, because of the steric

4.2 Results and Discussion 91

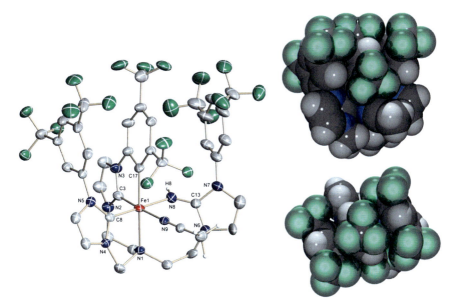

Fig. 4.6 Molecular structure of the complex cation of [(TIMEN^{3,5CF3}* = N^H)Fe(NCMe)](BPh$_4$) · solv (**5** · solv; solv = Et$_2$O · 1.43 MeCN) (30% probability ellipsoids). Hydrogen atoms (except H8, and hydrogens of coordinated acetonitrile), BPh$_4^-$ anion, and co-crystallized solvents are omitted for clarity. Space-filling representations of the cation **5** side view (*right, top*) and top view (*right, bottom*). Selected bond distances (Å): Fe1–N1 2.221(5), Fe1–N8 2.061(6), Fe1–N9 1.924(6), Fe1–C3 1.870(7), Fe1–C8 1.941(7), Fe1–C17 2.002(6), N8–C13 1.270(9)

demand of the CF$_3$-groups, the cavity formed by the tripodal ligand provides more space for the axial ligand (Fig. 4.4).

Aiming at a high-valent iron nitride complex photolysis of the azide complex with the light of a mercury vapour lamp was performed. During irradiation the orange acetonitrile solution turned deep brown. The resulting compound is NMR active und shows a diamagnetic spectrum. The new complex exhibits a Mößbauer isomer shift, δ of 0.27(1) mm s^{-1} and a quadrupole splitting, ΔE_Q, of 1.04(1) mm s^{-1}. The molecular structure of the cation [(TIMEN^{3,5CF3}* = N^H) Fe(NCMe)]$^+$ in brown crystals of [(TIMEN^{3,5CF3}* = N^H) Fe(NCMe)](BPh$_4$) · Et$_2$O · 6.5 MeCN (**5** · Et$_2$O · 6.5 MeCN), obtained from the solvent mixture Et$_2$O/MeCN, displays that upon photolysis several reaction steps must have occurred until the product is attained (Figs. 4.5 and 4.6).

The product is a pseudo octrahedral iron(II) complex, the iron center is coordinated by the inserted nitrogen atom, the cyclometallated as well as one unaltered carbene moiety and a solvent molecule, namely acetonitrile. Most probably, the first step of the insertion is the formation of an iron nitride complex. Trapping and characterization of an iron nitride complex during the photolysis could not be achieved, due to the highly reactive nature of the intermediate. As consecutive reaction we assume the cyclometallation along with a proton transfer from the aryl substiuent to the inserted nitride nitrogen. A diamagnetic NMR confirms the $S = 0$ groundstate of this molecule, in contrast to the usually observed high-spin Fe(II) complexes with trigonal pyramidal geometry. Complex **5** has a C_1 symmetry resulting in a large number and splitting of signals in the ^{1}H-NMR spectrum. For an isotope-labeling experiment the azide ligand in precursor **4** was labeled with terminal ^{15}N enriched sodium azide (Na$^+$ ^{15}N $=^{14}$N $=^{14}$N$^-$). In doing so, we could get a ^{15}N-NMR of the insertion product. The spectrum shows one expected peak at 320 ppm, referenced to liquid ammonia, which can be assigned to the imine nitrogen atom.

4.3 Conclusion

The *tris*-[2-(3-aryl-imidazol-2-ylidene)ethyl]amine ligand family (TIMEN^R, R = aryl = 2,6-xylyl (xyl), mesityl (mes)) has provided access to reactive transition metal complexes and the synthesis of an iron(IV) nitride complex was achieved. Derivatization of the TIMEN^R sytem (R = aryl = tolyl (tol), 3,5-xylyl (3,5xyl)) by utilizing sterically less demanding aryl substituents bearing potential reactive C–H bonds in *ortho* position of the aromatic rings, resulted in a dramatic change in reactivity of the Fe(II) precursor complexes. Aiming at Fe$\equiv$N complexes with enhanced reactivity another approach was carried out by introducing a 3,5-*bis*(trifluoromethyl)-phenyl substituent; thus, employing a compromise between steric bulk and accessibility towards the iron center. Here, a new tripodal N-heterocyclic carbene ligand of the TIMEN^R system (R = aryl = 3,5-(CF$_3$)aryl (3,5CF$_3$)), was synthesized. With this ligand, the Fe(II) precursor complex [(TIMEN^{3,5CF3})Fe(Cl)](Cl) (**3**), could be obtained, and via salt metathesis the corresponding azide complex [(TIMEN^{3,5CF3})Fe(N$_3$)](N$_3$) (**4**) was synthesized. The latter showed unexpected reactivity upon photolysis. Photolysis of **4** yielded [(TIMEN^{3,5CF3}* = N^H)Fe(NCMe)](BPh$_4$) (**5**), a complex, in which the nitride ligand inserted into one of the Fe–C bonds plus cyclometallation as seen in Chap. 3. All new metal complexes were characterized by single crystal X-ray structure determination. This fine-tuning of the TIMEN^R ligand system showed, once again, that relatively small changes in the periphery of the ligand have drastic effects on the reactivity of the resulting complexes. Significant differences in reactivity arise from the C–H bonds in *ortho* position of the aromatic rings. These potential reactive C–H bonds undergo aryl-cyclometallation induced by either reducing agents or photolysis. The increase of steric demand by introduction of trifluoromethyl instead of methyl groups in 3,5-position of the aryl substituents is insufficient to prevent C$_{aryl}$–H bond activation. As a result the TIMEN^{3,5CF3} ligand allows the synthesis of an iron(II) chloro complex which can be reacted further by adding an excess of sodium azide to yield the corresponding iron(II) azide complex. But it turned out that photolysis of the azide species does not deliver the desired iron(IV) nitride complex. Most probably the iron(IV) nitride species is a highly reactive interme-diate, which cannot be trapped. We assume that the nitride ligand inserts into one of

4.3 Conclusion

the iron–carbene bonds along with photoinduced cyclometallation of one of the aryl substituents. A subsequent hydrogen transfer from the aryl moiety to the inserted nitride could explain why this insertion reaction is more reproducible than the insertion reaction observed upon oxidation of [(TIMEN^{mes}) Fe^{IV}(N)] (BPh$_4$), as reported in Chap. 2. However, in order to get tripodal carbene complexes bearing sterically less demanding aryl substituents but, at the same time, allow an side access to the metal center, one possible solution could be the introduction of aryl groups with fluorinated 2,6-postions.

4.4 Experimental

4.4.1 Methods, Procedures, and Starting Materials

All air- and moisture-sensitive experiments were performed under dry nitrogen atmosphere using standard Schlenk line techniques or an MBraun inert-gas glovebox containing an atmosphere of purified dinitrogen. Anhydrous iron(II) chloride, 99.9%, was purchased from Aldrich and used as received. Sodium azide, sodium tetraphenylborate, potassium *tert*-butoxide, and sodium *tert*-butoxide were purchased from ACROS and were used without further purification. Solvents were purified using a two-column solid-state purification system (Glasscontour System, Irvine, CA) and transferred to the glove box without exposure to air. NMR solvents were obtained packaged under argon and stored over activated molecular sieves and sodium (where appropriate) prior to use. *Tris*(2-chloroethyl)amine, 3,5-[*bis*(trifluoromethyl)phenyl]imidazole were synthesized according to literature [10, 11]. Elemental analysis were performed at the Analytical Laboratories at the Friedrich-Alexander-University Erlangen-Nuremberg (Erlangen, Germany). Due to poor solubility, limited stability, and difficulties in separation of the reactants during work-up of some of the compounds, certain routine analytical methods could not be applied. Yields were calculated and noted where possible. All compounds were characterized by single crystal X-ray structure determination with crystals being representative for the bulk sample.

^{1}H-NMR spectra were recorded on JEOL 270 and 400 MHz instruments, operating at respective frequencies of 269.714 and 400.178 MHz with a probe temperature of 23 °C. ^{13}C-NMR spectra were recorded on JEOL 270 and 400 MHz instruments, operating at respective frequencies of 67.82 MHz and 100.624 MHz with a probe temperature of 23 °C. Chemical shifts were reported in ppm relative to the peak of SiMe$_4$ using ^{1}H (residual) chemical shifts of the solvent as a secondary standard. ^{15}N-NMR data were acquired on a JEOL 400 MHz instrument, and chemical shifts were referenced to CH$_3$NO$_2$ (380.23 ppm relative to liquid ammonia at 0 ppm). ^{57}Fe Mößbauer spectra were recorded on a WissEl Mößbauer spectrometer (MRG-500) at 77 K in constant acceleration mode. ^{57}Co/ Rh was used as the radiation source. WinNormos for Igor Pro software was used

94   4 TIMEN[3,5CF3]: Hybrid System Capable of Cyclometallation and Nitride Insertion

for the quantitative evaluation of the spectral parameters (least-squares fitting to Lorentzian peaks). The minimum experimental line widths were 0.23 mm s^{-1}. The temperature of the samples was controlled by an MBBC-HE0106 MÖSS-BAUER He/N$_2$ cryostat within an accuracy of ± 0.3 K. Isomer shifts were determined relative to α-iron at 298 K.

Samples for mass spectrometry were prepared by dissolving the compound in dry acetonitrile under nitrogen to yield a 10^{-4} M solution. Using a syringe pump at a flow rate of 240 mL h^{-1}, the acetonitrile solutions were infused into an orthogonal ESI source of an Esquire 6000 ion trap mass spectrometer (Bruker, Bremen, Germany). Nitrogen was used as the nebulizing gas. The source voltages for the positive-ion mode were set as follows: capillary at -4.0 kV, end plate offset at -500 V, capillary exit at 220.0 V. The ion trap was optimized for the respective target mass.

4.4.2 Synthetic Details

[H$_3$TIMEN^{3,5CF3}](Cl)$_3$ (1). A 100 mL flask was charged with *tris*(2-chloroethyl)amine (6.61 g, 27.42 mmol), 3,5-[*bis*(trifluoromethyl)phenyl]imidazole (10.00 g, 35.69 mmol) and dissolved in THF. The mixture was refluxed at 100 °C overnight, during which time an off-white precipitate formed. The hygroscopic chloride salt was collected by filtration, washed with diethyl ether and *n*-pentane, and dried *in vacuo* for two days at 70 °C (23.00 g 22.00 mmol, yield 88.80%). ^{1}H-NMR (270 MHz, RT, DMSO-d_6): δ [ppm] = 9.89 (s, 3 H), 7.93 (s, 3 H), 8.50 (s, 6 H), 8.45 (s, 3 H), 8.42 (s, 3 H), 4.40 (t, 6 H, ^3J = 6.4 Hz), 3.18 (t, 6 H, ^3J = 6,4 Hz). ^{13}C{^{1}H}-NMR (100 MHz, RT, DMSO-d_6): δ [ppm] = 214.93 (3C), 135.17 (C), 132.17 (3C), 130.52 (6C), 123.38 (3C), 121.32 (6C), 120.77 (3C), 51.32 (3C), 46.13 (3C), 20.57 (3C). Elemental analysis (%) for [H$_3$TIMEN^{3,5CF3}](Cl)$_3 \cdot$ H$_2$O, C$_{39}$H$_{32}$Cl$_3$F$_{18}$N$_7$O calcd. C 44.06, H 3.03, N 9.22; obsd. C 43.94, H 2.85, N 8.79.

[TIMEN^{3,5CF3}] (2). A solution of potassium *tert*-butoxide (216 mg, 1.90 mmol) in 10 mL diethyl ether was added to a suspension of [H$_3$TIMEN^{3,5CF3}](BF$_4$)$_3$ (750 mg, 0.63 mmol) in 10 mL of diethyl ether and stirred for 3 h. The solution was filtered through Celite and evaporated to dryness *in vacuo*. ^{1}H-NMR (270 MHz, RT, benzene-d_6): δ [ppm] = 8.08 (s, 6H), 7.49 (s, 3H), 6.49 (s, 3H), 6.36 (s, 3H), 3.83 (t, 6H, ^3J = 6.6 Hz), 2.73 (t, 6H, ^3J = 6.6 Hz), 2.00 (s, 9H). ^{13}C{^{1}H}-NMR (100.5 MHz, benzene-d_6, RT): δ [ppm] = 214.42 (3 C), 139.46 (q, 6 C, ^1J(^{13}C^{19}F) = 13,34 Hz), 68.09 (3 C), 49.62 (3 C).

[(TIMEN^{3,5CF3})Fe(Cl)](Cl) (3). A solution of [TIMEN^{3,5CF3}] (530 mg, 0.57 mmol) in 10 mL THF was added to a suspension of FeCl$_2$ (710 mg, 0.56 mmol) in 5 mL of THF and stirred overnight, during which time an off-white precipitate formed. The precipitate was collected by filtration, washed with diethyl ether and *n*-pentane, and dried in *vacuo* (513 mg, 0.48 mmol, yield 84.74%). ^{1}H-NMR (270 MHz, RT, acetonitrile-d_3): δ [ppm] = 113.07 (3H), 25.42 (3H), 24.40 (3H), 16.67 (3H), 11.80 (3H), 9.75 (3H), -3.20 (3H). Mößbauer (solid state, 77 K)

4.4 Experimental

Fig. 4.7 ^{1}H-NMR spectrum of [(TIMEN^{3,5CF3})Fe(Cl)](Cl) **3** (recorded in acetonitrile-d_3)

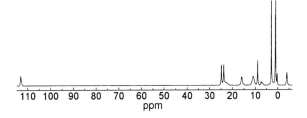

Fig. 4.8 ^{1}H-NMR spectrum of [TIMEN^{3,5CF3}Fe(N$_3$)](BPh$_4$) **(4)** (recorded in acetonitrile-d_3)

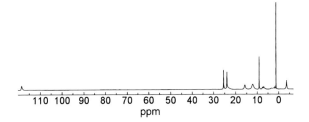

$\delta = 0.78(1)$ mm s^{-1}, $\Delta E_Q = 3.42(1)$ mm s^{-1}, $\Gamma_{FWHM} = 0.42(1)$ mm s^{-1}. Elemental analysis (%) for [(TIMEN^{3,5CF3})Fe(Cl)]Cl · CHCl$_3$, C$_{40}$H$_{28}$Cl$_5$F$_{18}$N$_7$ calcd. C 40.65, H 2.39, N 8.30; obsd. C 40.72, H 2.57, N 8.13. ESI–MS for C$_{39}$H$_{27}$FeClF$_{18}$N$_7$ calcd. 1026 [M$^+$]; obsd. 1026 [M$^+$] (Fig. 4.7).

[TIMEN^{3,5CF3}Fe(N$_3$)](BPh$_4$) (4). A suspension of [(TIMEN^{3,5CF3})Fe(Cl)](Cl) (100 mg, 0.09 mmol) and sodium azide (12.3 mg, 0.19 mmol) in acetonitrile was stirred overnight. The resulting orange solution was filtered through Celite and evaporated to dryness *in vacuo*. Orange crystals suitable for X-ray diffraction analysis were grown by diffusion of diethyl ether into a saturated solution of **4** in acetonitrile at room temperature. **4** was obtained by dissolving equivalent amounts of NaBPh$_4$ with [(TIMEN^{3,5CF3*})Fe(N$_3$)](N$_3$) in acetonitrile. The solution was filtered over Celite and directly used for crystallization attempts. IR (KBr): $\tilde{v}$ [cm^{-1}] = 2956 (w, CH$_2$), 2069 (s, N$_3$), 1278 (ss, CF$_3$), 1178 (m, CF$_3$), 1130 (s, CF$_3$). ^{1}H-NMR (acetonitrile-d_3): δ [ppm] = 118.19 (3H), 25.90 (3H), 24.46 (3H), 16.35 (3H), 12.75 (3H), 9.71 (3H), -2.91 (3H). Mößbauer (solid state, 77 K) δ = 0.80(1) mm s^{-1}, $\Delta E_Q = 2.90(1)$ mm s^{-1}, $\Gamma_{FWHM} = 0.61(1)$ mm s^{-1}. ESI–MS (m/z) for C$_{39}$H$_{27}$FeF$_{18}$N$_{18}$ calcd. 1033 [M$^+$]; obsd. 1033 [M$^+$] (Figs. 4.8 and 4.9).

[(TIMEN^{3,5CF3*} = N^H)Fe(MeCN)](BPh$_4$) (5). A stirred solution of [TIMEN^{3,5CF3}Fe(N$_3$)](N$_3$) (50 mg, 4.59 · 10^{-2} mmol) in 20 mL THF was irradiated for 24 h (mercury vapor lamp) during which the orange solution turned brown. The solution was filtered through Celite, concentrated and addition of *n*-pentane to the solution induced precipitation of a brown powder. The precipitate was collected, washed with *n*-pentane, and dried in a vacuum. Brown crystals suitable for X-ray diffraction analysis were grown by diffusion of diethyl ether into a saturated solution of 5 in acetonitrile at room temperature. **5** was obtained by dissolving equivalent amounts of NaBPh$_4$ with [(TIMEN^{3,5CF3*} = N^H)Fe(MeCN)](N$_3$) in acetonitrile. The solution was filtered over Celite and directly used for

96   4 TIMEN^{3,5CF3}: Hybrid System Capable of Cyclometallation and Nitride Insertion

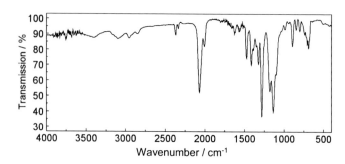

Fig. 4.9 IR spectrum of [TIMEN^{3,5CF3}Fe(N$_3$)](BPh$_4$) (**4**) (KBr pellet), azide band at 2069 cm^{-1}

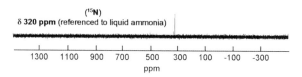

Fig. 4.10 ^{15}N-NMR spectrum of [(TIMEN^{3,5CF3*} = N^H)Fe(MeCN)](BPh$_4$) (**5**) (recorded in DMSO-d_4)

Fig. 4.11 ^{1}H-NMR spectrum of [(TIMEN^{3,5CF3*} = N^H)Fe(MeCN)](BPh$_4$) (**5**) (recorded in DMSO-d_6)

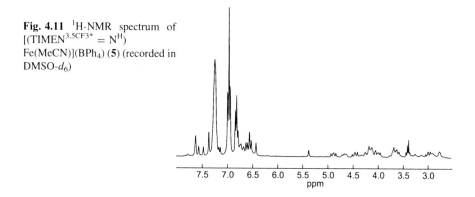

crystallization attempts. ^{15}N-NMR (400 MHz, DMSO-d_6, 20 °C): δ [ppm] = 320 (ref. to liquid ammonia). Mößbauer (solid state, 77 K) δ = 0.27(1) mm s^{-1}, ΔE_Q = 1.04(1) mm s^{-1}, Γ_{FWHM} = 0.53(1) mm s^{-1}. ESI–MS (m/z) for C$_{41}$H$_{30}$FeF$_{18}$N$_9$ calcd. 1047 [M$^+$]; obsd. 1047 [M$^+$] (Figs. 4.10 and 4.11).

4.4.3 X-ray Crystal Structure Determination Details

Greenish yellow prisms of [(TIMEN^{3,5CF3}) Fe(Cl)](Cl) · KCl · 2 Et$_2$O (**3** · KCl · 2 Et$_2$O) were obtained by slow diethyl ether diffusion into a saturated

4.4 Experimental

Table 4.2 Crystallographic data, data collection, and refinement details of [(TIMEN$^{3.5CF3}$)Fe(Cl)](Cl) · KCl · 2 Et$_2$O (**3** · KCl · 2 Et$_2$O), [(TIMEN$^{3.5CF3}$)Fe(N$_3$)](BPh$_4$) · Et$_2$O (**4** · Et$_2$O), and [(TIMEN$^{3.5CF3}$*= N^H)Fe(NCMe)](BPh$_4$) · solv (**5** · solv; solv = Et$_2$O · 1.43 MeCN)

	3 · KCl · 2 Et$_2$O [(TIMEN$^{3.5CF3}$)Fe(Cl)] (Cl) · KCl · 2 Et$_2$O	**4** · Et$_2$O [(TIMEN$^{3.5CF3}$)Fe(N$_3$)] (BPh$_4$) · Et$_2$O	**5** · solv [(TIMEN$^{3.5CF3}$*=N^H)] (BPh$_4$) · solv
Empirical formula	C$_{164}$H$_{128}$Cl$_9$F$_{72}$FeKN$_{28}$O$_2$	C$_{67}$H$_{57}$BF$_{18}$FeN$_{10}$O	C$_{71.86}$H$_{64.28}$BF$_{18}$FeN$_{10.43}$O
Mol. weight	4472.49	1426.89	1498.55
Crystal size [mm^3]	0.28 × 0.14 × 0.08	0.28 × 0.07 × 0.03	0.13 × 0.10 × 0.07
Temperature [K]	150	150	100
Crystal system	Triclinic	Monoclinic	Monoclinic
Space group	P $\bar{1}$ (no. 2)	P2$_1$/c (no. 14)	P2$_1$/c (no. 14)
a [Å]	20.172(2)	20.348(3)	12.7670(9)
b [Å]	22.609(3)	13.938(2)	12.3075(10)
c [Å]	23.730(3)	22.735(3)	43.956(3)
α [°]	62.30(1)	90	90
β [°]	80.106(6)	94.94(2)	90.284(3)
γ [°]	78.697(6)	90	90
V [Å^3]	9355(2)	6424(2)	6906.8(9)
Z	2	4	4
ρ_{calc} [g cm^{-3}]	1.588	1.475	1.441
μ [mm^{-1}]	0.586	0.340	0.320
F (000)	4496	2920	3078
T_{min}; T_{max}	0.796; 0.954	0.840; 0.990	0.567; 0.746
2θ interval [°]	6.5 ≤ 2θ ≤ 50.8	12.0 ≤ 2θ ≤ 40.0	6.62 ≤ 2θ ≤ 49.4
Collected reflections	213458	93686	49177
Independent reflections; R_{int}	34151; 0.2033	12156; 0.1293	11681; 0.0799
Observed reflections [$I \geq 2\sigma(I)$]	19251	6954	8364
No. refined parameters	2682	1070	1135
wR_2 (all data)	0.1788	0.1763	0.2467
R_1 [$I \geq 2\sigma(I)$]	0.0630	0.0657	0.1015
GooF on F^2	1.028	1.064	1.184
$\Delta\rho_{max/min}$	1.145/−0.895	0.836/−0.562	1.179/−0.517

Table 4.3 Selected bond distances [Å] and bond angles [°] with e.s.d.'s in parentheses for [(TIMEN^{3,5CF3}) Fe(Cl)](Cl) · KCl · 2 Et$_2$O (3 · KCl · 2 Et$_2$O), [(TIMEN^{3,5CF3}) Fe (N$_3$)](BPh$_4$) · Et$_2$O (4 · Et$_2$O)

	3 · KCl · 2 Et$_2$O	4 · Et$_2$O
Fe1–N1	3.084(4)	3.082(5)
Fe1–Lig$_{axial}$[a]	2.310(2)	1.966(4)
Fe1–C3	2.115(5)	2.116(4)
Fe1–C8	2.115(5)	2.109(4)
Fe1–C13	2.112(5)	2.122(4)
N8–N9	–	1.074(16)
N9–N10	–	1.171(17)
N2–C3	1.367(6)	1.354(5)
N3–C3	1.381(6)	1.379(5)
N4–C8	1.359(6)	1.358(5)
N5–C8	1.383(6)	1.377(5)
N6–C13	1.359(6)	1.365(5)
N7–C13	1.376(6)	1.368(5)
Fe1–N8–N9	–	144.0(2)
N8–N9–N10	–	175.0(2)
C3–Fe1–Lig$_{axial}$	100.9(2)	102.5(2)
C8–Fe1–Lig$_{axial}$[a]	105.9(2)	102.0(2)
C13–Fe1–Lig$_{axial}$[a]	104.0(2)	107.5(2)
$d_{Fe\ out\text{-}of\text{-}plane\ shift}$	0.510(3)	0.515(3)

[a] Lig$_{axial}$ corresponds to the respective axial Ligand, for compound **3** : Cl1 (chloro); for compound **4** : N8 (azide)

dichloromethane solution of the complex. Orange needles of [(TIME-N^{3,5CF3})Fe(N$_3$)](BPh$_4$) · Et$_2$O (**4** · Et$_2$O) were obtained by slow diethyl ether diffusion into an acetonitrile solution of the complex. Yellow plates of [(TIMEN^{3,5CF3}* = N^H) Fe(NCMe)] (BPh$_4$) · solv (**5** · solv; solv = Et$_2$O · 1.43 MeCN) were obtained by slow diethyl ether diffusion into an acetonitrile solution of the complex. Suitable single crystals of the compounds were embedded in protective perfluoropolyalkylether oil and quickly transferred to the cold nitrogen gas stream of the diffractometer. Intensity data were collected on a Bruker-Nonius KappaCCD diffractometer using graphite monochromatized MoK_α radiation (λ = 0.71073 Å). Lorentz and polarization effects were taken into account during data reduction, semiempirical absorption corrections were performed on the basis of multiple scans using [12]. All structures were solved by direct methods and refined by full-matrix least-squares procedures on F^2 using [13].

The asymmetric unit of [(TIMEN^{3,5CF3}) Fe(Cl)](Cl) · KCl · 2 Et$_2$O (**3** · KCl · 2 Et$_2$O) contains four independent molecules of the complex as well as one KCl and two molecules of diethyl ether. Some of the CF$_3$-groups of the ligand are disordered, two preferred orientations were refined with following occupancies: F13–F15/F13A–F15A = 0.70(2)/0.30(2); F207–F209/F227–F229 = 0.674(9)/0.326(9); C231,F210–F212/C251,F230–F232 = 0.631(7)/0.369(7); F307–F309/F327–F329 = 0.634(9)/0.366(9); F401–F403/F421–F423 = 0.55(3)/0.45(3). SIMU and SAME restraints were applied to some extent in the refinement of the disorder. Two of the chloride anions are disordered over three crystallographic positions, associated site

4.4 Experimental

Table 4.4 Selected bond distances [Å] and bond angles [°] with e.s.d.'s in parentheses for [(TIMEN$^{3,5CF3}_* = N^H$) Fe(NCMe)](BPh$_4$) · solv (**5** · solv; solv = Et$_2$O · 1.43 MeCN)

	5 · solv
Fe1–N1	2.221(5)
Fe1–N8	2.061(6)
Fe1–N9	1.924(6)
Fe1–C3	1.870(7)
Fe1–C8	1.941(7)
Fe1–C17	2.002(6)
N8–H8	0.8800
N8–C13	1.272(9)
N2–C3	1.343(8)
N3–C3	1.350(8)
N4–C8	1.372(8)
N5–C8	1.374(8)
N6–C13	1.365(8)
N7–C13	1.371(9)
C3–Fe1–N9	174.8(3)
C3–Fe1–C8	96.7(3)
N9–Fe1–C8	88.5(2)
C3–Fe1–C17	82.3(3)
N9–Fe1–C17	98.1(2)
C8–Fe1–C17	94.3(3)
C3–Fe1–N8	83.7(3)
N9–Fe1–N8	91.1(2)
C8–Fe1–N8	178.5(2)
C17–Fe1–N8	84.3(3)
C3–Fe1–N1	89.3(2)
N9–Fe1–N1	89.6(2)
C8–Fe1–N1	93.3(2)
C17–Fe1–N1	169.3(2)
N8–Fe1–N1	88.2(2)
C13–N8–Fe1	121.6(5)

occupancies factors were refined with following values: Cl8 = 0.649(7), Cl9 = 0.699(7) und Cl10 = 0.651(7).

[(TIMEN^{3,5CF3} Fe (N$_3$)] (BPh$_4$) · Et$_2$O (**4** · Et$_2$O) crystallizes with one molecule of diethyl ether per formula unit. The azid ligand is disordered. Two preferred orientations were refined each resulting in occupancies of 58(2) and 42(2) % for atoms N9–N10 and N9A–N10A, respectively. All CF$_3$-groups of the ligand are disordered, two preferred orientations were refined with following occupancies: F1–F3/F1A–F3A = 0.53(2)/0.47(2); F4–F6/F4A–F6A = 0.54(2)/0.46(2); F7–F9/F7A–F9A = 0.44(11)/0.56(11); F10–F12/F10A–F12A = 0.74(1)/0.26(1); F13–F15/F13A–F15A = 0.73(2)/0.27(2) und F16–F18/F16A–F18A = 0.23(2)/0.77(2). SIMU, ISOR and SAME restraints were applied in the refinement of the disorder. In the computations of the crystal structure determination of **3** · KCl · 2 Et$_2$O and **4** · Et$_2$O all non-hydrogen atoms were refined with anisotropic displacement parameters. All hydrogen atoms were placed in positions of optimized geometry, their isotropic displacement parameters were tied to those of the corresponding carrier atoms by a factor of either 1.2 or 1.5.

The asymmetric unit of [(TIMEN^{3,5CF3}* = N^H)Fe(NCMe)](BPh$_4$) · solv (**5** · solv; solv = Et$_2$O · 1.43 MeCN) contains four independent molecules of the complex as well as one molecule of diethyl ether and 1.43 molecules of MeCN. Five of the six CF$_3$ groups of the ligand are subjected to rotational disorder. Two alternative positions were refined in each case that were occupied by 78(2) and 22(2) % of the atoms F4–F6 and F4A–F6A, 78(1) and 22(1) % for the atoms F7–F9 and F7A–F9A, 69(3) and 31(3) % for F10–F12 and F10A–F12A, 77(2) and 23(2) % for F13–F15 and F13A–F15A and 65(2) and 35(2) for F16–F18 and F16A–F18A, respectively. The compound crystallizes with one molecule Et$_2$O that is disordered (refined occupancies of 57(2) and 43(2) % for the atoms C101–C105 and C120–C125) and 1.43 molecules of MeCN with 0.43 MeCN sharing its site with the site of the minor occupied Et$_2$O. SADI, SIMU, ISOR and SAME restraints were applied in the refinement of the disordered structure parts. In the computations of the crystal structure determination of **5** · solv all non-hydrogen atoms were refined with anisotropic displacement parameters. All hydrogen atoms were placed in positions of optimized geometry, their isotropic displacement parameters were tied to those of the corresponding carrier atoms by a factor of either 1.2 or 1.5 (Tables 4.2, 4.3, 4.4).

References

1. W.B. Tolman (ed.), *Activation of Small Molecules: Organometallic and Bioinorganic Perspectives* (Wiley-VCH, Weinheim, 2006)
2. X.L. Hu, I. Castro-Rodriguez, K. Olsen, K. Meyer, Organometallics **23**, 755 (2004)
3. M. Alcarazo, T. Stork, A. Anoop, W. Thiel, A. Fürstner, Angew. Chem. Int. Ed. **49**, 2542 (2010)
4. F.E. Hahn, M.C. Jahnke, Angew. Chem. Int. Ed. **47**, 3122 (2008)
5. C. Vogel, F.W. Heinemann, J. Sutter, C. Anthon, K. Meyer, Angew. Chem. Int. Ed. **47**, 2681 (2008)
6. J.J. Scepaniak, M.D. Fulton, R.P. Bontchev, E.N. Duesler, M.L. Kirk, J.M. Smith, J. Am. Chem. Soc. **130**, 10515 (2008)
7. X. Hu, K. Meyer, J. Am. Chem. Soc. **126**, 16322 (2004)
8. C.S. Vogel, F.W. Heinemann, M.M. Khusniyarov, K. Meyer, Inorg. Chim. Acta **364**, 226 (2010)
9. P. Gütlich, Chemie in unserer Zeit **4**, 133 (1970)
10. K. Ward Jr., J. Am. Chem. Soc. **57**, 914 (1935)
11. A. J. Arduengo III, F. P. Gentry Jr., P. K. Taverkere, H. E. Simmons III, (E. I. Du Pont de Nemours & Co., USA) 2001
12. Sadabs 2.06, Bruker AXS, Inc., 2002, Madison
13. Shelxtl NT 6.12, Bruker AXS, Inc., 2002, Madison

Chapter 5
PhB(^{tBu}Im)$_3^-$: Synthesis, Structure, and Reactivity of an Iron(V) Nitride

5.1 Introduction

High-valent iron species are proposed as active intermediates in the cycles of many important biocatalysts. Iron(IV) is the most readily accessible high oxidation state, although iron(V) has also been proposed as a key intermediate in some non-heme dioxygenases [1]. Structural and spectroscopic precedent for these inter-mediates has often come from studies of well-defined model complexes. For example, the iron(IV) oxo moiety, which has long been known to be at the cat-alytic centers of oxygenases [2], was first crystallographically characterized in an octahedral iron complex of an *N*-methylated 1,4,8,11-tetraazacyclotetradecane (cyclam) macrocycle [3]. Similar cyclam derivatives have also allowed for the preparation and detailed spectroscopic characterization of octahedral Fe(V) [4] and Fe(VI) [5] nitrido complexes, but the structural characterization and reactivity of these fleeting intermediates, which are usually studied at low temperatures in frozen matrices, remains elusive.

Iron nitrides have also been proposed to be key intermediates in the industrial (Haber–Bosch process) [6] and biological (nitrogenase) [7, 8] synthesis of ammonia. While iron bound "surface nitride" species are observed on the catalyst surface in the Haber–Bosch process [9], recent X-ray diffraction studies on nitrogenase suggest an interstitial atom in the center of the iron–sulfur cluster of the FeMo cofactor [10]. Although it is tempting to suggest a nitride anion at the site of biological nitrogen reduction, the nature of this atom is controversial and still under debate [11, 12]. Accordingly, the synthesis and characterization of model complexes is critical to delineating the reactivity of iron nitrides and their possible role in ammonia synthesis. In addition to the molecular and electronic structural insight provided by complexes that stabilize the [Fe≡N] unit, their reactivity will also impact our understanding of both biological and industrial NH$_3$ syntheses.

C. S. Vogel, *High- and Low-Valent tris-N-Heterocyclic Carbene Iron Complexes*, 101
Springer Theses, DOI: 10.1007/978-3-642-27254-7_5,
© Springer-Verlag Berlin Heidelberg 2012

5.2 Results and Discussion

Tripodal N-heterocyclic carbene ligands enable the synthesis of thermally stable, four-coordinate iron(IV) nitrido complexes as described in Chap. 2. The TIMEN^{mes} supported iron(IV) nitrido complex is stunningly air and moisture stable in solid state and at room temperature. Accordingly [(TIMEN^{mes})Fe(N)](BPh$_4$) shows low reactivity, thus being unsuitable for reactivity studies. Attempts to oxidize this iron(IV) compound in order to obtain an iron(V) species unexpectedly resulted in an iron(II) imine species that had undergone an insertion of the nitrido ligand into one of the iron–carbene bonds. The molecular flexibility of the TIMEN^{mes} ligand seems to be inappropriate in this context. Similarly, tripodal *tris*(phosphino)borato supporting ligands are capable of forming thermally stable, four-coordinate iron(IV) nitrido complexes, although these compounds could not be isolated in the solid state [13]. In contrast to the sterically demanding *tris*(carbene)amino system, the *tris*(phosphino)borato ligand can not protect sufficiently the highly reactive [Fe≡N] unit yielding a dinitrogen-bridged Fe(II)/Fe(II) complex resulting from dimerization. Using ligands that are structurally and electronically related to both *tris*(phosphino)borates and *tris*(carbene)amines, the molecular and electronic structures, as well as preliminary reactivity of low coordinate iron(IV) complexes that contain the [Fe≡N] functionality were recently reported [14, 15]. In this chapter the synthesis of a four-coordinate iron(V) nitrido complex supported by a tripodal N-heterocyclic carbene (NHC) ligand is described. For the first time an iron(V) nitride complex has been characterized by single crystal X-ray diffraction (XRD) [16]. Further characterization was carried out by Mößbauer and electron paramagnetic resonance (EPR) spectroscopies as well as density functional theory (DFT) calculations. The enhanced reactivity of this complex is evident from a low temperature hydrolysis reaction, which gives small amounts of ammonia, in contrast to the related iron(IV) nitrido complex, which is unreactive to water. Substantially higher yields of ammonia are obtained when the hydrolysis is conducted under reducing conditions, where water and the reductant provide three hydrogen atom equivalents. This reaction is reminiscent of the chemistry of nitrogenase, where water is the ultimate source of protons in the synthesis of ammonia.

The synthesis as well as the molecular and crystal structure of the four-coordinate nitrido complex [PhB(^{tBu}Im)$_3$Fe^{IV}≡N] (**1**), where PhB(^{tBu}Im)$_3^-$ is the phenyl*tris*(3-*tert*-butylimidazol-2-ylidene)borato ligand were recently reported [15]. The Mößbauer spectrum of **1** (Fig. 5.1) confirms the oxidation state assignment, with a negative isomer shift ($\delta = -0.28(1)$ mm s^{-1}) and large quadrupole splitting ($|\Delta E_Q| = 6.23(1)$ mm s^{-1}) similar to other iron(IV) nitrido complexes.

At slow scan rates, the cyclic voltammogram of complex **1** (Fig. 5.2) features a quasireversible oxidative wave that becomes fully reversible when the scan rate is increased ($E_{1/2} = -0.53$ V vs. the ferrocene/ferrocenium couple in tetrahydrofuran solvent with 0.4 M tetrabutylammonium hexafluorophosphate electrolyte).

5.2 Results and Discussion

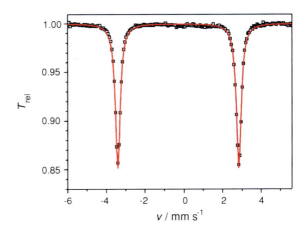

Fig. 5.1 Zero-field ^{57}Fe Mössbauer spectrum of a microcrystalline sample of **1** recorded at 77 K. The *solid line in red* represents the best fit obtained with parameters: $\delta = -0.28(1)$ mm s^{-1}, $|\Delta E_Q| = 6.23(1)$ mm s^{-1}

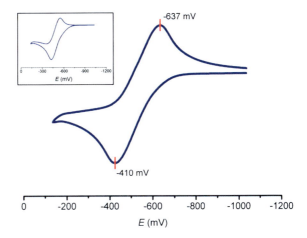

Fig. 5.2 Cyclic voltammogram of **1**, showing the Fe^V/Fe^{IV} couple, 0.4 M NBu$_4$PF$_6$ in THF, 22 °C. Scan rate 200 mV s^{-1}. *Inset*: 20 mV s^{-1}

The reversibility of the Fe^V/Fe^{IV} redox couple at faster scan rates suggests that an oxidized iron nitrido complex is kinetically accessible.

In agreement with the electrochemical data, **1** is readily oxidized by [Fe(Cp)$_2$]BAr$_{F24}$ (Cp$^-$ = η^5-C$_5$H$_5^-$; BAr$_{F24}^-$ = B(3,5-(CF$_3$)$_2$C$_6$H$_3$)$_4^-$) at −78 °C in diethyl ether solution to generate the dark purple complex [PhB(^{tBu}Im)$_3$-Fe^V≡N]BAr$_{F24}$ in high yield (**2**, Scheme 5.1). No resonances are observed in the ^{1}H-NMR spectrum of complex **2**, indicating complete consumption of the diamagnetic starting material. The UV-vis spectrum of **2** (Fig. 5.3) is distinct from that of **1** (Fig. 5.4), with bands at 452 nm (ε = 1500 M^{-1} cm^{-1}) and 563 nm (ε = 1400 M^{-1} cm^{-1}), which are assigned as ligand to metal charge transfer transitions on the basis of time-dependent DFT calculations. The half-life of **2** is approximately 4 h at 25 °C as determined by UV-vis spectroscopy and we therefore manipulated the complex at low temperatures.

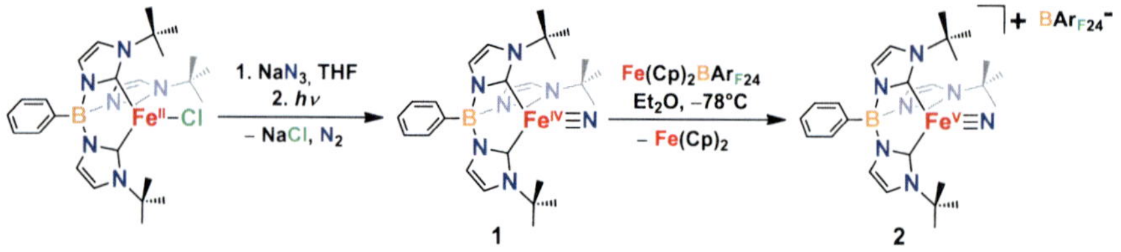

Scheme 5.1 Formation of Fe(V) nitride **2**

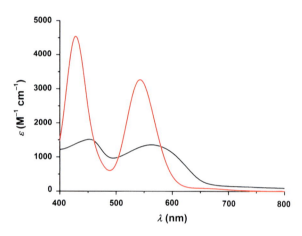

Fig. 5.3 Experimental UV-vis spectrum of **2** recorded in Et$_2$O at 22 °C (*black line*) and calculated UV-vis spectrum of **2** as obtained from the spin-unrestricted B3LYP-TD-DFT calculations (*red line*). Full-width at half-maximum (FWHM) height was set to 2000 cm^{-1} for each transition

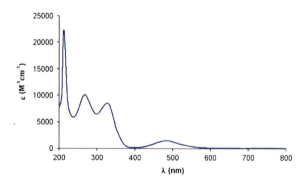

Fig. 5.4 Experimental UV-vis spectrum of **1** recorded in Et$_2$O at 22 °C

Purple crystals suitable for X-ray crystal structure determination were grown by slow diffusion of *n*-pentane into a toluene solution of **2** at −35 °C. The molecular structure of **2** shows a pseudo-tetrahedral iron center that is supported by the *tris*(imidazol-2-ylidene)borato ligand and bound to a terminal nitrido ligand. The nitrido ligand lies in a hydrophobic pocket created by the bulky *tert*-butyl groups of the *tris*(imidazol-2-ylidene)borato ligand (Fig. 5.5).

The crystallographic data, collected at 100 and 35 K on two single crystals obtained from independent syntheses reveal that there are remarkably few

5.2 Results and Discussion

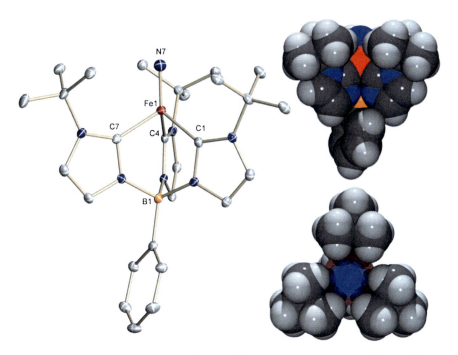

Fig. 5.5 Molecular structure of [PhB(^{tBu}Im)$_3$Fe^V≡N]$^+$ in crystals of [PhB(^{tBu}Im)$_3$Fe^V≡N]-BAr$_{F24}^-$ · 2.5 toluene · 0.5 *n*-pentane (**2** · 2.5 toluene · 0.5 *n*-pentane) at 35 K (left, 50% probability ellipsoids). Hydrogen atoms, co-crystallized solvent molecules and the BAr$_{F24}^-$ counter anion are omitted for clarity. Space-filling representations of the cation **2** side view (*right, top*) and top view (*right, bottom*). Selected bond distances (Å) and angles (°): Fe1–N7 1.506(2), Fe1–C1 1.947(2), Fe1–C4 1.969(2), Fe1–C7 1.932(2), Fe1···B1 2.949(3), N7–Fe1–C1 119.62(10), N7–Fe1–C4 126.47(10), N7–Fe1–C7 114.84(9), C1–Fe1–C4 98.16(9), C1–Fe1–C7 93.67(9), C4–Fe1–C7 97.44(9), N7–Fe1···B1 173.57(8)

structural changes upon oxidation of the iron(IV) complex. In both nitrido complexes, the iron center is four-coordinate and lies approximately 1 Å out of the plane defined by the three carbon atoms of the *tris*(imidazol-2-ylidene)borato ligand. The one-electron oxidation Fe(IV) → Fe(V) is accompanied by a slight shortening of the Fe–N bond from 1.512(1) Å in Fe(IV) to 1.506(2) and 1.502(2) Å found in the two independent XRD studies at 35 and 100 K, respectively. This decrease can be readily explained as the outcome of increased electrostatic interactions between the more positively charged Fe(V) ion and the N^{3-} nitrido ligand. However, reasons for the concomitant and significant elongations of the three Fe–C bonds, from 1.924(1) Å (mean value in Fe(IV)) to 1.949(2) and 1.947(2) Å (mean values in the Fe(V) complex at 35 and 100 K), are much less obvious. The σ-donating carbenes are formally neutral ligands that experience a much smaller, if any, electrostatic attraction to the highly oxidized Fe(V) ion. The carbenes interact mostly with the iron $d(xz)$, $d(yz)$ and $d(z^2)$ orbitals, which are the same orbitals involved in bonding to the nitrido ligand. The shorter Fe–N distance

Table 5.1 Reduced orbital charges for **1** and **2** (spin-unrestricted B3LYP-DFT calculations, Löwdin populations)

		1: Fe(IV)	2: Fe(V)	Δ, Fe(V)–Fe(IV)
Fe	s	6.376	6.388	0.012
	$d(z^2)$	1.175	1.231	0.055
	$d(xz)$	1.239	1.251	0.012
	$d(yz)$	1.245	1.274	0.028
	$d(x^2-y^2)$	1.640	1.659	0.019
	$d(xy)$	1.630	1.354	−0.276
Nitride	s	3.640	3.646	0.006
	p_z	1.082	1.085	0.002
	p_x	1.265	1.190	−0.075
	p_y	1.274	1.179	−0.095

increases the Fe–N orbital overlap, increasing the electron density at these d-orbitals (see Löwdin populations, Table 5.1), reducing the Fe–C orbital overlap, and thus increasing the Fe–C bond lengths. Finally, the B1–Fe1–N7 angle decreases upon oxidation from $178.57(6)°$ in **1** to $173.57(8)°$ in **2**. Noticeable deviation of this angle from $180°$ in **2** is related to the electronic structure *(vide infra)*.

The $Fe \equiv N$ bond lengths in both **1** and **2** are significantly shorter than the $Fe \equiv N$ bond of 1.57 ± 0.02 Å in Fe(VI) nitride $[(Me_3\text{-cyclam-ac})Fe \equiv N]^{2+}$, determined by X-ray absorption spectroscopy (or the DFT computed distance of 1.532 Å) [5]. This difference is likely due to the higher coordination number of the six-coordinate iron center in hexavalent $[(Me_3\text{-cyclam-ac})Fe \equiv N]^{2+}$, which is expected to result in a longer $Fe(VI) \equiv N$ triple bond than the corresponding distance in the four-coordinate $Fe(V) \equiv N$ complex **2**. Additionally, it would be reasonable to assume that the acetato ligand *trans* to the nitrido weakens the $Fe \equiv N$ interaction in $[(Me_3\text{-cyclam-ac})Fe \equiv N]^{2+}$. However, EXAFS studies on structurally related $[(Me_4\text{-cyclam})(X)Fe = O]^+$ complexes (X = OH^-, $CF_3CO_2^-$, N_3^-, NCS^-, NCO^-, and CN^-), show little change in the $Fe = O$ bond length within this series and with the parent acetonitrile complex [17]. In this regard, it is also worth mentioning that the Mößbauer isomer shift $\delta = -0.29$ mm s^{-1}, observed for $[(Me_3\text{-cyclam-ac})Fe \equiv N]^{2+}$, is considerably less negative than that measured for **2** ($\delta = -0.45$ mm s^{-1}, *vide infra*). In fact, the value determined for the Fe(VI) nitrido complex is more comparable to those of known Fe(IV) nitrido complexes, for which δ varies from -0.27 to -0.34 mm s^{-1}. Because the Mößbauer isomer shift is an indicator of formal oxidation state and also the degree of covalency, the more negative isomer shift in **2** suggests a stronger, more covalent Fe–nitride interaction, and thus, a shorter Fe–N bond distance in four-coordinate high-valent $Fe \equiv N$ complexes.

The electronic structure of **2** was investigated by temperature-dependent ^{57}Fe Mößbauer and X-band EPR spectroscopies. The zero-field ^{57}Fe Mößbauer spectrum, recorded at 78 K, reveals one major doublet, fitted with two quadrupole lines

5.2 Results and Discussion

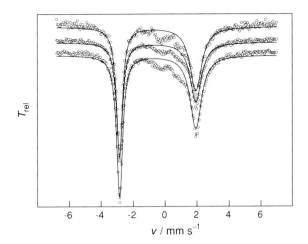

Fig. 5.6 Zero-field ^{57}Fe Mößbauer spectra of multiple independently synthesized microcrystalline samples of **2** recorded at 77 K

of different linewidths but equal integrated areas, which is characterized by a very low $\delta = -0.45$ mm s^{-1} and a large $|\Delta E_Q| = 4.78$ mm s^{-1}. While the low energy quadrupole line of this doublet is relatively sharp, the high energy line is considerably broader and shows additional features on its low-energy flank, possibly due to an unidentified and EPR-inactive Fe(II) or Fe(III) side product. It is interesting to note, however, that the intensity of these features is invariable for multiple independently synthesized samples (Fig. 5.6). Due to slow relaxation at low temperatures, the observation of broad lines and partly resolved magnetic hyperfine interactions in the Mößbauer spectrum is not unusual and frequently observed in spectra of systems with half-integer spin [18]. Accordingly, a spectrum recorded at 200 K shows a more symmetrical quadrupole doublet with significantly sharper lines (Fig. 5.7). The isomer shift decreases at the higher temperature, as expected from a second-order Doppler shift [19, 20]. The isomer shift of the parent Fe(IV) nitrido (**1**, low-spin, $S = 0$) is -0.28 mm s^{-1} (Fig. 5.1). Thus, significant lowering of the isomer shift upon oxidation **1** → **2** suggests a metal-based Fe(IV) → Fe(V) oxidation. An Fe(V) ion (d^3 electronic configuration) can be in either a low-spin (one unpaired electron, $S = 1/2$) or high-spin state (three unpaired electrons, $S = 3/2$), which is best probed by EPR spectroscopy (*vide infra*).

Wieghardt and co-workers reported the spectroscopic characterization of Fe(V)-nitride complexes with cyclam-derived ligands [21]. These six-coordinate complexes, which were originally formulated as high-spin ($S = 3/2$) [21] and then reformulated as low-spin ($S = 1/2$) [4], show considerably higher isomer shifts (-0.04 to -0.02 mm s^{-1}) and much smaller quadrupole splittings ($1.67 - 1.90$ mm s^{-1}) than complex **2**. Though the spin state of these species is still being debated, the significantly more negative isomer shift of **2** is due to increased covalent bonding in the *tris*(imidazol-2-ylidene)borato complex **2** and resulting decreased *d*-electron density compared with the less covalent Fe–N(cyclam) bonding. The profoundly different values of the quadrupole splitting for **2** and the

Fig. 5.7 Zero-field ^{57}Fe Mößbauer spectra of a microcrystalline sample of **2** recorded at 78 and 200 K. The *solid lines* represent the best fit obtained with parameters. At 78 K: $\delta = -0.45(1)$ mm s^{-1}, $|\Delta E_Q| = 4.78(1)$ mm s^{-1}, $\Gamma_{FWHM} = 0.44(1)/0.90(1)$ mm s^{-1}; at 200 K: $\delta = -0.49(1)$ mm s^{-1}, $|\Delta E_Q| = 4.73(1)$ mm s^{-1}, $\Gamma_{FWHM} = 0.28(1)/0.36(1)$ mm s^{-1}

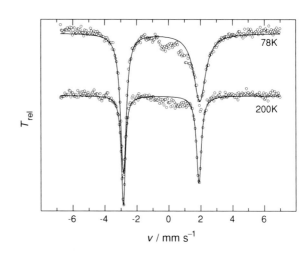

Fig. 5.8 X-Band EPR spectrum of **2** in frozen toluene measured at 15 K. Conditions: frequency 8.9564 GHz, power 1 mW, modulation 1 mT/100 kHz. The fit (*dashed line*) was obtained with $S = 1/2$ and $g_\perp = 1.971$, $g_\parallel = 2.299$

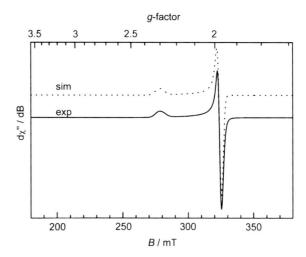

cyclam complexes are also a consequence of the different geometries. Recently, Collins and co-workers reported Mößbauer parameters very similar to **2** for a putative five-coordinate low-spin Fe(V) oxo complex [22] ($\delta = -0.42$ mm s^{-1}, $\Delta E_Q = +4.25$ mm s^{-1}), emphasizing similarities between the electronic structures of these two species.

The X-band EPR spectrum of **2** reveals an axial signal with $g_\perp = 1.971$ and $g_\parallel = 2.299$ (Fig. 5.8).

Due to the large anisotropy ($\Delta g = g_\parallel - g_\perp = 0.328$) and location of both signal features near $g = 2$, this spectrum is assigned to an iron-centered radical ($S = 1/2$) resulting from a low-spin Fe(V) configuration in the ground state of **2**. The observed axial EPR signal is expected for the approximately C_3-symmetric coordination environment of the metal ion in complex **2**. Note that an EPR signal with two components at $g \sim 4$ and one component at $g \sim 2$ would be expected

5.2 Results and Discussion

Table 5.2 Principal bond distances (Å) and an angle (°) of the optimized structure as obtained from the spin-unrestricted BP-DFT calculations

	2
Fe1–C1	1.941 (1.947(2))[a]
Fe1–C4	1.941 (1.969(2))
Fe1–C7	1.921 (1.932(2))
Fe1–N7	1.515 (1.506(2))
Fe1–B1	2.924 (2.949(3))
N7–Fe1–B1	171.60 (173.57(8))

[a] The experimental values are given in parentheses, 35 K data

for a high-spin Fe(V) complex ($S = 3/2$) of small rhombicity. Collins and co-workers reported a similar EPR spectrum for their five-coordinate low-spin Fe(V)=O complex, which shows a slightly distorted axial signal with lower g-values and the opposite order of $g_\parallel$ and $g_\perp$ components (1.74, 1.97, and 1.99) [22]. The electronic structure of **2** has been further elucidated by DFT calculations (B3LYP). The geometry of the complex was optimized as a spin doublet ($S = 1/2$), with all the geometrical features of **2** well-reproduced (Table 5.2). It is remarkable that the calculated Fe1–N7 bond length of 1.515 Å is very close to the experimental distance of 1.506 Å.

The optimized structure also reveals a slight deviation from a pseudo C_3-symmetry with a N7–Fe1–B1 angle of 171.6° similar to the experimentally determined angle of 173.6°. According to the Löwdin population analysis, complex **2** is an iron-centred radical: the spin-density at the metal ion is +0.93 (Fig. 5.9a). The unoccupied $d(z^2)$, $d(xz)$, $d(yz)$ orbitals are highly destabilized by strong σ- and π-interactions with nitrogen p-orbitals of the nitrido ligand (Fig. 5.9b). A singly occupied molecular orbital (SOMO) reveals predominant $d(xy)$ character, whereas the $d(x^2-y^2)$ highest occupied molecular orbital (HOMO) is doubly occupied. The N7–Fe1–B1 bending removes the degeneracy of the triply occupied $d(xy)$ and $d(x^2-y^2)$ orbitals, and thus, decreases the total energy. This bonding situation is very similar to diamagnetic four-coordinate Fe(IV)nitrido complexes with a pseudo C_3 symmetry [13 – 15], where $d(x^2-y^2)$ and $d(xy)$ are doubly occupied, and where ΔE_Q has been determined to be +6.23 mm s^{-1}. The calculated Mößbauer parameters for **2** ($\delta = -0.56$ mm s^{-1}, $\Delta E_Q = +4.63$ mm s^{-1}) are in good agreement with the experimental values ($\delta = -0.49$ mm s^{-1}, $|\Delta E_Q| = 4.73$ mm s^{-1} at 200 K), which corroborates the calculated electronic structure. Although we were able to calculate EPR parameters as well, only qualitative agreement between theory and experiment is reached. The calculated g-tensor (2.007, 2.013, 2.133) is almost axial in agreement with the experimental data but the spin-orbit coupling is underestimated by the calculations leading to a smaller anisotropy of the tensor.

Preliminary reactivity studies reveal **2** to be highly reactive, yielding substantial amounts of NH_3 under extremely mild conditions. Specifically, **2** reacts within seconds at –78 °C (in tetrahydrofuran solution) with three equivalents of cobaltocene ($[Co(Cp)_2]$) and fifteen equivalents of water to produce NH_3 in 89% yield, as determined by the indophenol method [23]. The "indophenol method" is

110   5 PhB(^{tBu}Im)$_3^-$: Synthesis, Structure, and Reactivity

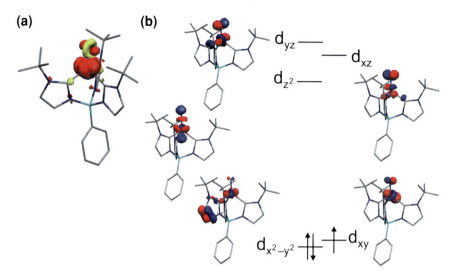

Fig. 5.9 Spin density map for **2** obtained from the spin-unrestricted B3LYP-DFT calculations (**a**) and frontier iron-based molecular orbitals of **2**: canonical β-orbitals are shown, the z-axis is parallel to the Fe≡N vector (**b**)

Table 5.3 Yields of ammonia from the reaction of cobaltocene and water (pH 7), as determined by the indophenol method

Entry	Equivalents [Co(Cp)$_2$]	Equivalents H$_2$O	NH$_3$% yield
1	1	15	13.0
2	2	15	47.9
3	3	5	77.8
4	3	15	88.6
5	3	15	81.5
6	3	15	81.7
7	3	30	73.5

Table 5.4 Yields of ammonia at different pH values

pH	Yield of NH$_3$ (%)
5.1	11.8
7.1	12.2
9.1	10.7

particularly sensitive to nitrogen determination from ammonia, in which NH$_3$ is oxidized in an alkaline hypochlorite solution to NH$_2$Cl, which, in the presence of sodium nitroprusside as a catalyst, subsequently reacts with a phenol (thymol) to yield the deep blue, spectro-photometrically determined indophenol. Fewer than three equivalents of reductant resulted in lower yields of ammonia. Mößbauer spectroscopic analysis reproducibly revealed a single iron(II) product, consistent with a three-electron reaction (for details see Tables 5.3 and 5.4 and Fig. 5.10).

While the mechanism of this reaction is likely to be complicated, we note that protonation of the nitrido ligand following one-electron reduction to the Fe(IV)

Table 5.5 Crystallographic data, data collection, and refinement details of [PhB(^{tBu}Im)$_3$Fe^V(N)]BAr$_{F24}$ at 35 and 100 K

	[PhB(^{tBu}Im)$_3$Fe^V≡N]BAr$_{F24}$@35 K	[PhB(^{tBu}Im)$_3$Fe^V≡N]BAr$_{F24}$@100 K
Empirical formula	C$_{79}$H$_{76}$B$_2$F$_{24}$FeN$_7$	C$_{78.95}$H$_{76.10}$B$_2$F$_{24}$FeN$_7$
Mol. weight	1656.94	1656.44
Crystal size [mm^3]	0.49 × 0.20 × 0.18	0.24 × 0.20 × 0.12
Temperature [K]	35	100
Crystal system	Monoclinic	Monoclinic
Space group	$P2_1/c$ (no. 14)	$P2_1/c$ (no. 14)
a [Å]	18.474(2)	18.6042(9)
b [Å]	23.619(2)	23.698(1)
c [Å]	19.385(2)	19.488(1)
α [°]	90	90
β [°]	111.714(1)	111.783(1)
γ [°]	90	90
V [Å^3]	7858(2)	7978.4(7)
Z	4	4
ρ_{calc} [g cm^{-3}]	1.401	1.379
μ [mm^{-1}]	0.297	0.292
F (000)	3404	3403
T_{min}; T_{max}	0.603; 0.746	0.659; 0.746
2θ interval [°]	$4.12 \leq 2\theta \leq 54.2$	$4.1 \leq 2\theta \leq 54.2$
Collected reflections	60563	70131
Independent reflections; R_{int}	17330;0.0617	17596;0.0329
Observed reflections [$I \geq 2\sigma(I)$]	11946	14144
Refined parameters	1171	1199
wR_2 (all data)	0.1073	0.1113
R_1 [$I \geq 2\sigma(I)$]	0.0427	0.0437
GooF on F^2	1.022	1.022
$\Delta\rho_{max/min}$	0.465/−0.649	0.587/−0.475

Fig. 5.10 Zero-field ^{57}Fe Mößbauer spectra of solutions resulting from reaction of **2** with cobaltocene/water (6×10^{-2} M and 3×10^{-1} M, with naturally abundant ^{57}Fe). Spectra were obtained for two independent experiments and recorded at 77 K

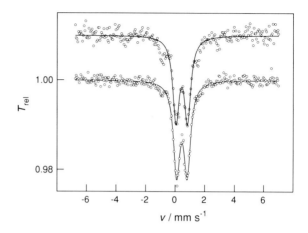

nitrido complex **1** is unlikely, since independent experiments show that **1** does not react with water. The rapid formation of NH$_3$ at low temperature, with H$_2$O serving as the hydrogen atom source, highlights the highly reactive nature of the [Fe^V≡N] unit in **2**.

5.3 Conclusion

Despite being implicated as important intermediates, iron(V) compounds have proven very challenging to isolate and characterize. The preparation of the iron(V) nitrido complex, [PhB(^{tBu}Im)$_3$Fe^V≡N]BAr$_{F24}$ (PhB(^{tBu}Im)$_3^-$ = phenyl*tris*(3-*tert*-butyl-imidazol-2-ylidene)borato, BAr$_{F24}$ = B(3,5-(CF$_3$)$_2$C$_6$H$_3$)$_4^-$) by one electron oxidation of the iron(IV) nitrido precursor is discussed in this chapter. Single crystal X-ray diffraction of the iron(V) complex reveals a four-coordinate metal ion with a terminal nitrido ligand. Mößbauer and electron paramagnetic resonance spectroscopic characterization, supported by electronic structure calculations, provide evidence for a d^3 iron(V) metal center in a low spin ($S = 1/2$) electron configuration. Low temperature reaction of the iron(V) nitrido complex with water under reducing conditions leads to high yields of ammonia with concomitant formation of an Fe(II) species.

5.4 Experimental

5.4.1 Methods, Procedures, and Starting Materials

All manipulations were performed under a nitrogen atmosphere by standard Schlenk techniques or in an M. Braun Labmaster glove box maintained at or below 1 ppm of O$_2$ and H$_2$O. Glassware was dried at 150 °C overnight. Diethyl ether,

5.4 Experimental

n-pentane and toluene were purified by the Glass Contour solvent purification system. The complexes $[PhB(^{tBu}Im)_3Fe^{IV} \equiv N]$ (**1**) [15] and $[Fe(Cp_2)]BAr_{F24}$ [24] were prepared according to literature procedures. UV-vis spectra were recorded with a CARY 100 Bio UV-visible spectrophotometer. Cyclic voltammograms were recorded with a Bioanalytical Systems CV-50 W voltammetric analyzer, a platinum wire counter electrode, a silver wire reference electrode and a platinum disk working electrode. Ferrocene was used as an internal standard. EPR spectra were recorded on a JEOL CW spectrometer JESFA200 equipped with an X-band Gunn diode oscillator bridge, a cylindrical mode cavity, and a helium cryostat. ^{57}Fe Mößbauer spectra were recorded on a WissEl Mößbauer spectrometer (MRG-500) at 77 K in constant acceleration mode. The temperature of the samples was controlled by an MBBC-HE0106 MÖSSBAUER He/N_2 cryostat within an accuracy of ± 0.3 K. Isomer shifts were determined relative to α-iron at 298 K.

5.4.2 Computational Details

The program package ORCA 2.7 revision 0 was used for all calculations [25]. The geometry optimization calculations were performed by the spin-unrestricted DFT method with the BP86 functional [26–28]. The single point calculations and calculations of Mößbauer parameters were performed with the B3LYP functional. The triple-ζ basis sets with one-set of polarization functions (TZVP) [29] were used for iron ions and the double-ζ basis sets with one-set of polarization functions (SVP) [29] were used for all other atoms. For calculation of Mößbauer parameters, the "core" CP(PPP) basis set for iron [30, 31] was used [32]. TD-DFT calculations were performed using the B3LYP functional and the conductor like screening model (COSMO) [33] with diethyl ether as a solvent. The first 30 states were calculated, where the maximum dimension of the expansion space in the Davidson procedure (MAXDIM) was set to 300. The EPR g-tensor was calculated with the B3LYP functional, including relativistic effects in zero order regular approximation (ZORA) [34]. Molecular orbitals and the spin density plot were visualized via the program Molekel [35].

5.4.3 Synthetic Details

$[PhB(^{tBu}Im)_3Fe^{V} \equiv N]BAr_{F24}$ (**2**). A precooled solution of $[PhB(^{tBu}Im)_3Fe^{IV} \equiv N]$ (**1**) (51.5 mg, 96.9 μmol) in diethyl ether (5 mL) was mixed with a precooled solution of $[Fe(Cp_2)]BAr_{F24}$ (101.7 mg, 96.9 μmol) in diethyl ether (5 mL) at -78 °C. The resulting purple solution was stirred at -78 °C for 20 min, then the deeply colored product precipitated with cold n-pentane. The yellow solution of $[Fe(Cp)_2]$ was decanted and the product washed with cold n-pentane (2 × 5 mL). The purple solid was dried in vacuo (130.0 mg, 96%). The complex is thermally

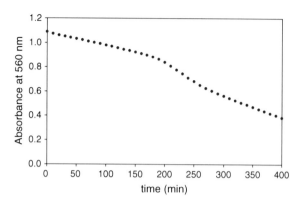

Fig. 5.11 Decomposition of **2** at 25 °C as measured by decay in the absorbance maximum at 560 nm

unstable and was always kept at temperatures below −35 °C. Complex **2** decays with a half-life of approximately 4 h at room temperature. Decomposition does not follow simple first-order kinetics (Fig. 5.11).

Formation of ammonia from 2, water and cobaltocene. A solution of cobaltocene (3 equivalents) and deionized, degassed water (see Table 5.3) in THF (ca. 1.5 mL) was cooled to −78 °C. In a separate flask, **1** (ca. 40 μmol) in Et$_2$O/THF (1:1, 5 mL) was oxidized at −78 °C with [Fe(Cp)$_2$]BAr$_{F24}$ to yield **2**. After stirring for 20 min (by which time oxidation to **2** is complete), the [Co(Cp)$_2$]/H$_2$O solution was added to the solution of **2**. Within 20 s, the color of the reaction mixture changed from purple to light brown. After stirring for 10 min at −78 °C, the solution was warmed to RT. A solution of HCl in diethyl ether (1 M, 1 mL) was injected into the flask and the solvents removed in vacuo (Note that control experiments showed that no ammonia was formed when **2** was treated with HCl). The ammonium chloride was extracted with deionized water (20 mL) and diluted to 25 mL in a volumetric flask. A 1.00 mL aliquot of the solution was used to quantify the ammonium chloride by the indophenol method [36, 37]. The yields of ammonia are collected in Table 5.3. Similar experiments conducted with water in the absence of the reductant gave small yields of ammonia (Table 5.4).

The formation of ammonia was confirmed in a separate experiment: following addition of the [Co(Cp)$_2$]/H$_2$O solution, the volatiles were vacuum transferred into a frozen solution of HCl in diethyl ether (1 M, 2 mL). The solution was thawed and stirred at room temperature for 30 min. The reaction was then dried in vacuo to yield a colorless solid, which was identified as NH$_4$Cl by ^{1}H-NMR spectroscopy (DMSO-d$_6$) [13].

After treatment of solutions of **2** (6 × 10^{-2} M and 3 × 10^{-1} M, with naturally abundant ^{57}Fe) with [Co(Cp)$_2$] in a THF/water mixture, the resulting solution reproducibly shows one Mößbauer quadrupole doublet with an isomer shift of 0.46(1) mm s^{-1} and a quadrupole splitting of 0.72(1) mm s^{-1} (Fig. 5.10). Attempts to isolate this new complex from the NMR silent solution have so far been unsuccessful, and solutions prepared for crystallization experiments decompose over time. Although the precise nature of the newly formed complex is not known at this time,

5.4 Experimental

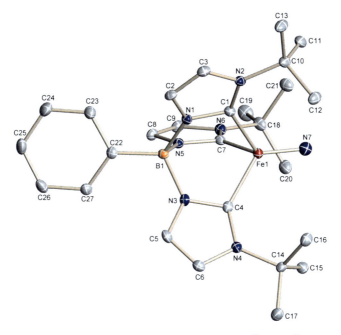

Fig. 5.12 Thermal ellipsoid plot of the complex cation of [PhB(^{tBu}Im)$_3$Fe^V≡N]BAr$_{F24}$ · 2.5 · toluene · 0.5 n-pentane @35 K (**2** · 2.5 · toluene · 0.5 n-pentane) with labeling scheme (hydrogen atoms omitted for clarity)

we propose the formation of an Fe(II) species, possibly [PhB(^{tBu}Im)$_3$Fe–OH$_2$]$^+$ or [PhB(^{tBu}Im)$_3$Fe–THF]$^+$, under these conditions. In an independent experiment, precursor **1** was treated with triphenylphosphine, Ph$_3$P, to yield the Fe(II) complex [PhB(^{Mes}Im)$_3$Fe–NPPh$_3$]. This compound shows remarkably similar Mößbauer parameters ($\delta = 0.44(1)$ mm s^{-1}, $\Delta E_Q = 0.78$ mm s^{-1}, at 300 K) compared to the above-described and prepared reaction mixture of **2**. The chemistry and electronic structure of [PhB(^{Mes}Im)$_3$Fe–NPPh$_3$] will be reported in Chap. 6.

5.4.4 X-ray Crystal Structure Determination Details

Purple blocks of [PhB(^{tBu}Im)$_3$Fe^V≡N]BAr$_{F24}$ were obtained by slow diffusion of n-pentane into a saturated toluene solution of the complex at −35 °C. Suitable single crystals of the compound were embedded in protective perfluoropolyalkylether oil and quickly transferred to the cold gas stream of the diffractometer. Intensity data on two different crystals from independent crystallizations were collected at 35 K on a Bruker Smart APEX2 diffractometer (MoK$_\alpha$ radiation, $\lambda = 0.71073$ Å, graphite monochromator) and at 100 K on a Bruker Kappa APEX2 Duo diffractometer equipped with an IμS microsource and QUAZAR focusing Montel optics (MoK$_\alpha$ radiation, $\lambda = 0.71073$ Å). Lorentz and

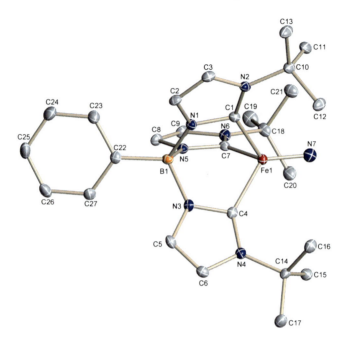

Fig. 5.13 Thermal ellipsoid plot of the complex cation of [PhB(^{tBu}Im)$_3$Fe^V≡N]BAr$_{F24}$ 2.48 · toluene · 0.52 · n-pentane @100 K (2 · 2.48 · toluene · 0.52 · n-pentane) with labeling scheme (hydrogen atoms omitted for clarity)

polarization effects were taken into account during data reduction, semi-empirical absorption corrections were performed on the basis of multiple scans using *SADABS* [38]. All structures were solved by direct methods and refined by full-matrix least-squares procedures on F^2 using *SHELXTL NT* 6.12 [39].

The unit cell of [PhB(^{tBu}Im)$_3$Fe^V≡N]BAr$_{F24}$@35 K contained a total of 2.5 molecules of toluene and 0.5 n-pentane with the partially occupied toluene and n-pentane sharing one crystallographic site. Three of the CF$_3$ groups of the BAr$_{F24}$ counterion are disordered. Two alternative orientations were refined in each case resulting in site occupancies of 71.7(7) and 28.3(7) % for F1–F3 and F1A–F3A, 68(2) and 32(2) % for F13–F15 and F13A–F15A and of 90(2) and 10(2) % for F16–F18 and F16A–F18A, respectively (Fig. 5.12).

The unit cell of [PhB(^{tBu}Im)$_3$Fe^V≡N]BArF$_{24}$@100 K contained a total of 2.48 molecules of toluene and 0.52 n-pentane sharing one crystallographic site with refined occupancies of 47.6(5) % for toluene (C301–C307) and 52.4(5) % for n-pentane (C311–C315). Four of the CF$_3$ groups of the counterion are disordered. Two alternative orientations were refined in each case resulting in site occupancies of 63.2(4) and 36.8(4) % for F1–F3 and F1A–F3A, 84.7(5) and 15.3(5) % for F4–F6 and F4A–F6A, 68.0(9) and 32.0(9) % for F13–F15 and F13A–F15A and of 72.9(7) and 27.1(7) % for F16–F18 and F16A–F18A, respectively. ISOR, SIMU, SADI and SAME restraints were applied in the refinement of the disordered parts of both structure determinations (Fig. 5.13).

5.4 Experimental

Acknowledgments Text, schemes, and figures of this chapter, in part, are reprints of the materials published in [16]. The dissertation author was the secondary researcher and author. The co-authors listed in the publication also participated in the research. The permission to reproduce the paper was granted by the American Association for the Advancement of Science. Copyright 2011, American Association for the Advancement of Science.

References

1. K. Chen, M. Costas, J. Kim, A.K. Tipton, L. Que, J. Am. Chem. Soc. **124**, 3026 (2002)
2. M. Costas, M.P. Mehn, M.P. Jensen, L. Que Jr., Chem. Rev. **104**, 939 (2004)
3. J.-U. Rohde, J.-H. In, M.H. Lim, W.W. Brennessel, M.R. Bukowski, A. Stubna, E. Münck, W. Nam, L. Que, Science **299**, 1037 (2003)
4. M. Aliaga-Alcalde, S.D. George, B. Mienert, E. Bill, K. Wieghardt, F. Neese, Angew. Chem. **117**, 2968 (2005)
5. J.F. Berry, E. Bill, E. Bothe, S.D. George, B. Mienert, F. Neese, K. Wieghardt, Science **312**, 1937 (2006)
6. G. Ertl, Chem. Rec. **1**, 33 (2001)
7. B.M. Hoffman, D.R. Dean, L.C. Seefeldt, Acc. Chem. Res. **42**, 609 (2009)
8. M.P. Hendrich, W. Gunderson, R.K. Behan, M.T. Green, M.P. Mehn, T.A. Betley, C.C. Lu, J.C. Peters, Proc. Natl. Acad. Sci. **103**, 17107 (2006)
9. G. Ertl, Angew. Chem. Int. Ed. **29**, 1219 (1990)
10. O. Einsle, F.A. Tezcan, S.L.A. Andrade, B. Schmid, M. Yoshida, J.B. Howard, D.C. Rees, Science **297**, 1696 (2002)
11. T.-C. Yang, N.K. Maeser, M. Laryukhin, H.-I. Lee, D.R. Dean, L.C. Seefeldt, B.M. Hoffman, J. Am. Chem. Soc. **127**, 12804 (2005)
12. D. Lukoyanov, V. Pelmenschikov, N. Maeser, M. Laryukhin, T.-C. Yang, L. Noodleman, D.R. Dean, D.A. Case, L.C. Seefeldt, B.M. Hoffman, Inorg. Chem. **46**, 11437 (2007)
13. T.A. Betley, J.C. Peters, J. Am. Chem. Soc. **126**, 6252 (2004)
14. C. Vogel, F.W. Heinemann, J. Sutter, C. Anthon, K. Meyer, Angew. Chem. Int. Ed. **47**, 2681 (2008)
15. J.J. Scepaniak, M.D. Fulton, R.P. Bontchev, E.N. Duesler, M.L. Kirk, J.M. Smith, J. Am. Chem. Soc. **130**, 10515 (2008)
16. J.J. Scepaniak, C.S. Vogel, M.M. Khusniyarov, F.W. Heinemann, K. Meyer, J.M. Smith, Science **331**, 1049 (2011)
17. T.A. Jackson, J.-U. Rohde, M.S. Seo, C.V. Sastri, R. DeHont, A. Stubna, T. Ohta, T. Kitagawa, E. Münck, W. Nam, J. Lawrence Que, J. Am. Chem. Soc, **130**,12394 (2008)
18. C.E. Schulz, P. Nyman, P.G. Debrunner, J. Chem. Phys. **87**, 5077 (1987)
19. R.H. Herber, *Chemical Mössbauer Spectroscopy* (Plenum Press, New York, 1984)
20. A.A. Yousif, H. Winkler, H. Toftlund, A.X. Trautwein, R.H. Herber, J. Phys.: Condens. Matter **1**, 7103 (1989)
21. K. Meyer, E. Bill, B. Mienert, T. Weyhermüller, K. Wieghardt, J. Am. Chem. Soc. **121**, 4859 (1999)
22. F.T. de Oliveira, A. Chanda, D. Banerjee, X. Shan, S. Mondal, L. Que Jr., E.L. Bominaar, E. Münck, T.J. Collins, Science **315**, 835 (2007)
23. M. Berthelot, Rep. Chim. Appl. **1**, 284 (1859)
24. J.L. Bras, H. Jiao, W.E. Meyer, F. Hampel, J.A. Gladysz, J. Organomet. Chem. **616**, 54 (2000)
25. F. Neese, *ORCA—an Ab Initio, Density Functional and Semiempirical SCF-MO Package, version 2.7 revision 0* (Institut für Physikalische und Theoretische Chemie Universität Bonn, Germany, 2009)
26. A.D. Becke, Phys. Rev. A **38**, 3098 (1988)

27. J.P. Perdew, Phys. Rev. B **34**, 7406 (1986)
28. A.D. Becke, J. Chem. Phys. **98**, 5648 (1993)
29. A. Schäfer, H. Horn, R. Ahlrichs, J. Chem. Phys. **97**, 2571 (1992)
30. S. Sinnecker, L.D. Slep, E. Bill, F. Neese, Inorg. Chem. **44**, 2245 (2005)
31. F. Neese, Inorg. Chim. Acta **337**, 181 (2002)
32. This basis is based on the TurboMole DZ basis, developed by Ahlrichs and coworkers and obtained from the basis set library under ftp://chemie.uni-karlsruhe.de/pub/basen
33. A. Klamt, G. Schürmann, *J Chem Soc, Perkin Trans* **799** (1993)
34. C. van Wüllen, J. Chem. Phys. **109**, 392 (1998)
35. S. Portmann, *Molekel, version 4.3.win32* (CSCS/UNI Geneva, Switzerland, 2002)
36. A.L. Chaney, E.P. Marbach, Clin. Chem. **8**, 130 (1962)
37. M.W. Weatherburn, Anal. Chem. **39**, 971 (1967)
38. SADABS 2008/1; Bruker AXS, Inc. (2008)
39. SHELXTL NT 6.12; Bruker AXS, Inc. (2002)

Chapter 6
PhB(^{Mes}Im)$_3^-$: Spin Crossover in a Four-Coordinate Iron(II) Complex

6.1 Introduction

Spin crossover complexes are characterized by the ability of a transition metal center to undergo a change in electronic configuration in response to external inputs such as heat, light, pressure or changes in magnetic field [1]. Spin crossover is one of the oldest molecular switching phenomena known, and perhaps the most vital, considering its central role in haemoglobin-based respiration. The phenomenon, first described in the 1930s, has since been observed for many complexes especially of iron (Fe^{II}, Fe^{III}) and cobalt, both in solution and in the crystalline state [2–4]. Since the molecular bistability of spin-crossover molecules is associated with a lack of fatigue, there is considerable interest for applications such as sensors and digital memory [5]. Spin crossover is typically observed in first-row transition metal complexes having d^5–d^7 electron counts. The largest group of spin crossover complexes is comprised of mononuclear iron(II) complexes with an FeN_6 coordination environment [6]. In these complexes, the $S = 0$ (t_{2g}^6)–S=2 ($t_{2g}^4 e_g^2$) transition is generally accompanied by a ca. 0.1 Å increase in Fe–N bond lengths. While a small number of higher coordinate spin crossover complexes are known, lower coordinate iron(II) complexes invariably assume a high-spin electronic configuration [7]. Indeed, reports of spin crossover in complexes featuring lower coordination numbers are so far limited to a small number of four-coordinate cobalt(II) complexes [8, 9]. In this chapter the first example of a four-coordinate iron(II) complex is described that undergoes spin crossover [10]. The transition from $S = 0$ to $S = 2$ has been characterized by dc magnetic susceptibility and Mößbauer spectroscopy. In addition, variable temperature single crystal X-ray diffraction has provided insights into the structural changes associated with the spin transition. Moreover, the electronic structures of the two spin states have been determined by density functional theory. Finally, both the *tris*(carbene)borato and phosphinamide ligands lend themselves to facile synthetic

---

C. S. Vogel, *High- and Low-Valent* tris-*N-Heterocyclic Carbene Iron Complexes*,
Springer Theses, DOI: 10.1007/978-3-642-27254-7_6,
© Springer-Verlag Berlin Heidelberg 2012

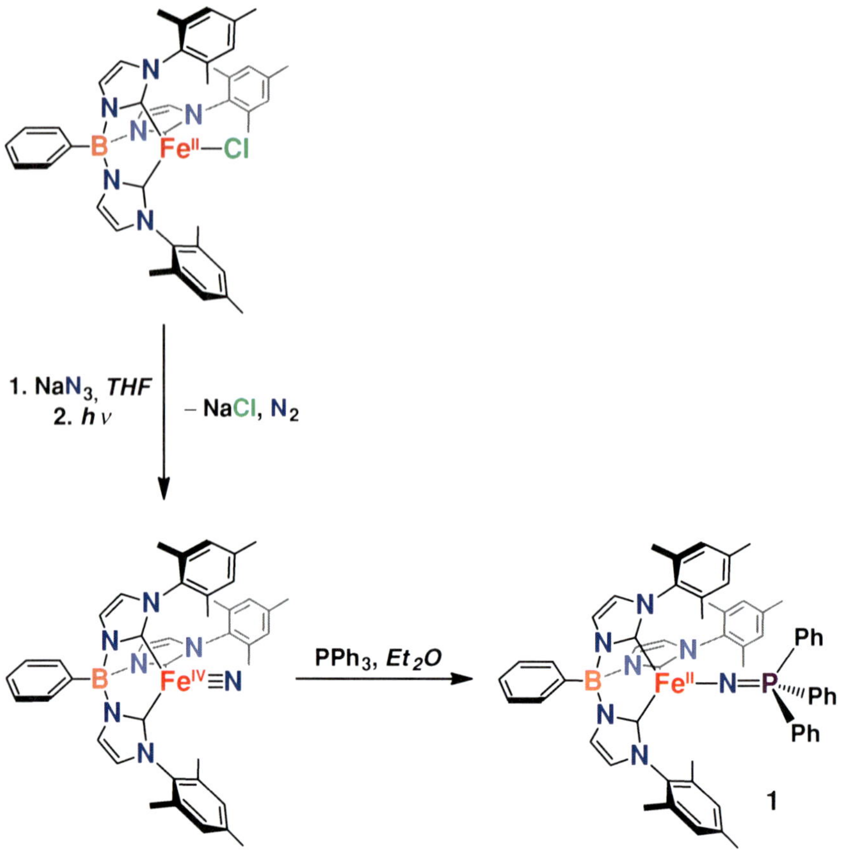

Scheme 6.1 Synthesis of [PhB(^{Mes}Im)$_3$Fe–N=PPh$_3$] (**1**)

modification, thus highlighting the potential of this system for tuneable spin crossover behavior.

6.2 Results and Discussion

Reaction of the iron(IV) nitrido complex, [PhB(^{Mes}Im)$_3$Fe≡N] [11], with PPh$_3$ results in the formation of yellow-green [PhB(^{Mes}Im)$_3$Fe–N=PPh$_3$] (6.1) in high yield (Scheme 6.1), similarly to a previous report on the synthesis of [PhB(^{tBu}Im)$_3$Fe–N=PPh$_3$] [12].

The complex has been structurally characterized at multiple temperatures (Figs. 6.4, 6.5). In solution at ambient temperature, **1** has a high-spin electronic configuration ($S = 2$), characterized by a paramagnetically shifted ^{1}H-NMR spectrum and a magnetic moment in solution of 5.2(3) μ_B.

6.2 Results and Discussion

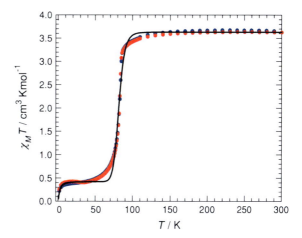

Fig. 6.1 Variable-temperature dc susceptibility data, collected for **1** while warming (*red*) and cooling (*blue*) under an applied field of 1000 Oe. The *solid black line* corresponds to a fit to the data, as described in the text

The solid-state magnetic behavior of **1** was probed by variable temperature dc susceptibility on a polycrystalline sample under an applied field of 1 T. As shown in Fig. 6.1, at 300 K, $\chi_M T = 3.63$ cm^3 mol K^{-1}, consistent with an $S = 2$ state with $g = 2.2$. There is little change in $\chi_M T$ as the temperature is decreased to 150 K, at which point $\chi_M T$ starts to gradually decrease before dropping precipitously at 81 K. Below 50 K $\chi_M T$ decreases slightly (from 10 to 50 K, average $\chi_M T = 0.42$ cm^3 mol K^{-1}), followed by a rapid decrease below 10 K, which is likely due to Zeeman and zerofield splitting. The $\chi_M T$ curve is essentially identical when the temperature is increased from 2 to 300 K, and notably, no hysteresis is evident at the applied sweep-rates. The precipitous drop in $\chi_M T$ at 81 K is indicative of a transition from a thermally excited high-spin $S = 2$ state at high temperature to a low-spin $S = 0$ ground state at low temperature. The average value of $\chi_M T = 0.42$ cm^3 mol K^{-1} below 50 K corresponds to 12% of the sample retaining a high-spin configuration at low temperature. Modeling the $\chi_M T$ data according to a Boltzmann distribution of spin states provides values of $\Delta H = 1190$ cm^{-1} and $T_C = 81$ K [13]. In addition, modeling the data below 20 K using MAGPACK [14] provides an axial zero-field splitting parameter of $|D| = 5(2)$ cm^{-1}.

The spin state changes have been confirmed by variable temperature zero-field Mößbauer spectroscopy (Fig. 6.2). At 7 K, a solid sample of **1** shows one major quadrupole doublet with an isomer shift of $\delta = -0.03(1)$ mm s^{-1} and a quadrupole splitting of $\Delta E_Q = 1.28(1)$ mm s^{-1} (93%). In addition, a minor feature, which integrates to approximately 7%, is observed as a shoulder on the high energy flank of the main doublet. This latter signal grows in as the temperature is increased from 7 to 40, 50, and 60 K. At 78 K, two sharp, equally intense quadrupole doublets with $\delta = -0.03(1)$ mm s^{-1}, $\Delta E_Q = 1.25(1)$ mm s^{-1} and $\delta = 0.55(1)$ mm s^{-1}, $\Delta E_Q = 1.17(1)$ mm s^{-1} are observed. This latter doublet continues to grow in as the temperature is gradually increased to 90, 100, and

Fig. 6.2 Variable temperature zero-field Mößbauer spectra for **1**

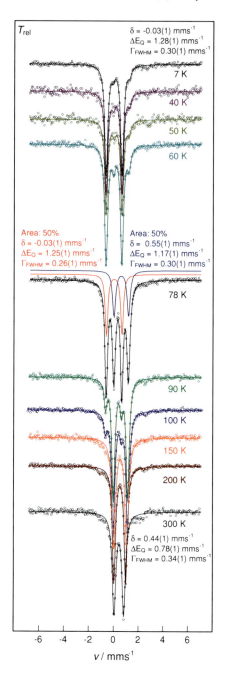

150 K, where the spectrum shows one doublet only [$\delta = 0.53(1)$ mm s^{-1}, $\Delta E_Q = 1.05(1)$ mm s^{-1}], characteristic of a high spin Fe(II) [PhB(^{Mes}Im)$_3$FeX] species (Fig. 6.3, Table 6.1).

6.2 Results and Discussion

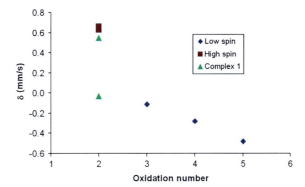

Fig. 6.3 Isomer shifts as a function of oxidation state for iron *tris*(carbene)borato complexes, [PhB(^{R}Im)$_3$FeX], R=^{t}Bu, Mes

In agreement with the dc susceptibility data, this doublet is assigned to the high spin state of **1**, with Mößbauer parameters of $\delta = 0.44(1)$ mm s^{-1} and $\Delta E_Q = 0.78(1)$ mm s^{-1} at 300 K. The structural variation associated with the spin state change has been established by X-ray crystallography (Figs. 6.4, 6.5).

The solid-state structure of **1** at 150 K, where the high-spin state is fully populated, shows similar metrical parameters to other *tris*(carbene)borato iron(II) complexes. For example, the Fe–C bond lengths (average 2.085(2) Å) are similar to the related iron(II) complex [PhB(^{Mes}Im)$_3$FeCl] [11]. It is notable that the Fe–N bond in **1** is slightly shorter than in the closely related high-spin complex [PhB(^{tBu}Im)$_3$Fe–N = PPh$_3$] [12], while the N–P bond is slightly longer. This difference is presumably a consequence of the different steric properties of the two *tris*(carbene)borato ligands. The solid-state structure of **1** was also determined at 30 K, where mostly the low-spin, $S = 0$ ground state is populated. At this temperature, the Fe–C bond lengths are over 0.1 Å shorter than in the high-temperature structure. In addition, there is a smaller, but significant reduction in the Fe–N bond distance (0.05 Å), along with a less pronounced lengthening of the N–P bond (0.03 Å). While the Fe–N–P bond angle is greater at 30 K, the possibility that solid-state packing forces are responsible for this change cannot be discounted. Two different Fe–N–P bond angles [167.8(1) and 177.6(1)°] are observed in the solid state structure of the high spin complex [PhB(^{tBu}Im)$_3$Fe–N = PPh$_3$] suggesting a low barrier to bending about this linkage [12]. The observed structural changes can be understood in terms of an orbital correlation diagram (Fig. 6.6a).

In idealized C_{3v} symmetry, the metal *d*-orbitals of **1** transform as $1a_1 + 2e$, similarly to other pseudotetrahedral complexes. The a_1 and $e_{(a)}$ sets of orbitals are non-bonding, while the $e_{(b)}$ orbitals have σ^* interactions with the supporting *tris*(carbene)borato ligand and π^* interactions with the axial phosphinamido ligand [18, 19]. The $S = 0$ to $S = 2$ spin transition therefore transfers two electrons from nonbonding to antibonding orbitals, resulting in longer Fe–C and Fe–N bonds. The dissimilar antibonding interactions result in greater elongation of the Fe–C bonds (σ^*) than the Fe–N bond (π^*). We note that the low-spin state is related to the four-coordinate *tris*(phosphino)borato iron(II) imido complex [PhB(CH$_2$PPh$_2$)$_3$Fe≡NAd]$^-$, which also has an $S = 0$ ground state [20]. Electronic structure

Table 6.1 Mößbauer spectral parameters for selected iron *tris*(carbene)borato complexes, recorded at 77 K

	Complex [Ref]	Oxidation number	S	δ (mm s^{-1})	ΔE_Q (mm s^{-1})
High spin	[PhB(^{Mes}Im)$_3$FeCl] [11]	2	2	0.66	1.83
	[PhB(^{Mes}Im)$_3$FeN$_3$] [11]	2	2	0.63	1.94
Low spin	[PhB(^{Mes}Im)$_3$Fe≡NAd] [15]	3	0.5	−0.11	1.65
	[PhB(^{Mes}Im)$_3$Fe≡N] [11]	4	0	−0.28	6.08
	[PhB(^{t}BuIm)$_3$Fe≡N] [16]	4	0	−0.28	6.23
	[PhB(^{t}BuIm)$_3$Fe≡N]$^+$ [17]	5	0.5	−0.45	4.78

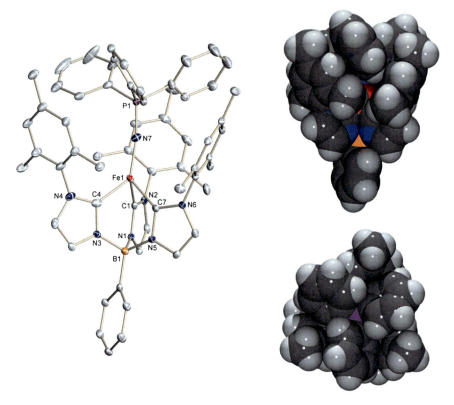

Fig. 6.4 Molecular structure of **1** at 30 K. Thermal ellipsoids shown at 50% probability, hydrogen atoms omitted for clarity. Space-filling representations of the cation **1** side view (*right, top*) and top view (*right, bottom*)

calculations are consistent with the qualitative orbital picture. Geometry optimization of the two spin states using DFT methods [BPW91/6-31G(d)//6-311G(d,f)] provides structures in reasonable agreement with the crystallographic data. The electronic structures were also calculated using the B3LYP functionals, but the BPW91 functional predicted the correct ground state. Importantly, the calculations reproduce the observed structural changes, with the longer Fe–N and Fe–C bonds

6.2 Results and Discussion

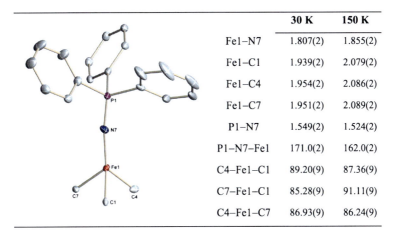

	30 K	150 K
Fe1–N7	1.807(2)	1.855(2)
Fe1–C1	1.939(2)	2.079(2)
Fe1–C4	1.954(2)	2.086(2)
Fe1–C7	1.951(2)	2.089(2)
P1–N7	1.549(2)	1.524(2)
P1–N7–Fe1	171.0(2)	162.0(2)
C4–Fe1–C1	89.20(9)	87.36(9)
C7–Fe1–C1	85.28(9)	91.11(9)
C4–Fe1–C7	86.93(9)	86.24(9)

Fig. 6.5 *Left* structural representation of the core structure of **1** at 30 K, showing selected bond lengths (Å) and bond angles (°). Thermal ellipsoids shown at 50% probability, hydrogen atoms and most of the *tris*(carbene)borato ligand omitted for clarity. *Right*. Comparison of structural parameters for **1** at 30 and 150 K

in the high spin state, and a longer N–P bond in the low spin state. Furthermore, the nature of the frontier orbitals agrees with the qualitative MO description. Thus, for $S = 0$ the HOMO has a_1 symmetry (Fig. 6.6b), while for $S = 2$, HOMO has e symmetry (Fig. 6.6c). An additional insight from these calculations is that the shorter P–N bond length in the high spin state is due to the P–N π-bonding contribution to the $e_{(b)}$ orbitals.

6.3 Conclusion

In summary, the first example of spin crossover in a four coordinate iron(II) complex has been characterized. A combination of SQUID magnetometry, Mößbauer spectroscopy, X-ray crystallography and electronic structure theory provide unambiguous evidence for a metal-based spin transition with $T_C = 81$ K. The four-coordinate iron(II) phosphinamido complex, [PhB(^{Mes}Im)$_3$Fe–N=PPh$_3$], undergoes an $S = 0 \rightarrow S = 2$ spin transition with $T_C = 81$ K, as determined by variable temperature magnetic measurements and Mößbauer spectroscopy. Variable temperature single-crystal X-ray diffraction reveals that the $S = 0 \rightarrow S = 2$ transition is associated with an increase in the Fe–C and Fe–N bond distances and a decrease in the N–P bond distance. These structural changes have been interpreted in terms of electronic structure theory. It is likely that the robust *tris*(carbene)borato ligand platform will stabilize other iron(II) spin crossover complexes that can be accessed by substitutions of the axial ligand. Moreover, the electronic structure predicts that the spin crossover temperature can be modulated by changes to the π-acceptor properties of the axial ligand.

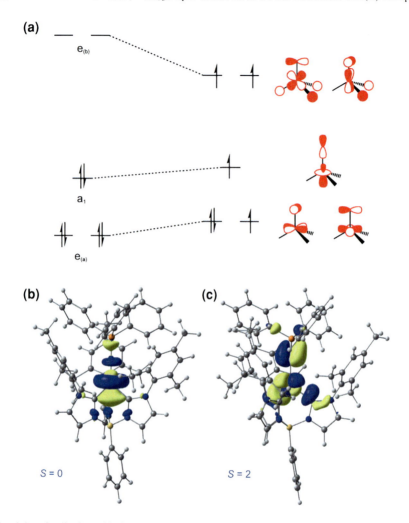

Fig. 6.6 a Qualitative orbital correlation diagram for the $S = 0$ to $S = 2$ transition in **1**; highest occupied molecular orbitals as calculated by DFT methods for **b** $S = 0$ and **c** $S = 2$ spin states

6.4 Experimental

6.4.1 Methods, Procedures, and Starting Materials

All manipulations were performed under a nitrogen atmosphere by standard Schlenk techniques or in an M. Braun Labmaster glove box maintained at or below 1 ppm of O_2 and H_2O. Glassware was dried at 150 °C overnight. ^{1}H-NMR data were recorded on a Varian Unity 400 spectrometer (400 MHz) at 22 °C. Diethyl ether, pentane, tetrahydrofuran and toluene were purified by the Glass Contour

6.4 Experimental

solvent purification system. Deuterated benzene was first dried over CaH_2, then over Na/benzophenone, and then vacuum transferred into a storage container. Before use, an aliquot of each solvent was tested with a drop of sodium benzophenone ketyl in THF solution. Celite was dried overnight at 200 °C under vacuum. [PhB(^{Mes}Im)$_3$Fe≡N] [11] was prepared according to a literature procedure. Triphenylphosphine was obtained from Strem Chemicals and recrystallized twice from diethyl ether before use. All other chemicals were obtained commercially and used as received. All signals in the ^{1}H-NMR spectra are referenced to residual C_6D_5H at δ 7.16 ppm.

Solution magnetic susceptibilities were determined by Evans' method [21].

UV–vis spectra were recorded with a CARY 100 Bio UV–visible spectrophotometer.

Elemental analysis was performed by Robertson Microlit Laboratories, Madison NJ.

^{57}Fe Mößbauer spectra were recorded on a WissEl Mößbauer spectrometer (MRG-500) at 77 K in constant acceleration mode. ^{57}Co/Rh was used as the radiation source. WinNormos for Igor Pro software was used for the quantitative evaluation of the spectral parameters (least-squares fitting to Lorentzian peaks). The minimum experimental line widths were 0.23 mm s^{-1}.

The temperature of the samples was controlled by an MBBC-HE0106 MÖSSBAUER He/N$_2$ cryostat within an accuracy of ±0.3 K. Isomer shifts were determined relative to α-iron at 298 K.

Magnetic data were collected using a Quantum Design MPMS-XL SQUID magnetometer. Measurements were obtained for a finely ground microcrystalline powder restrained within a polycarbonate gel capsule. Dc susceptibility data were collected in the temperature range 2–300 K under a dc field of 1 T. The data were corrected for core diamagnetism of the sample, estimated using Pascal's constants.

6.4.2 Synthetic Details

[PhB(^{Mes}Im)$_3$Fe-N = PPh$_3$] (1). Solutions of [PhB(^{Mes}Im)$_3$Fe≡N] (29 mg, 40.6 μmol) in diethyl ether (1.5 mL) and PPh$_3$ (11 mg, 41.9 μmol) in diethyl ether (1.5 mL) were combined at RT. A yellow-green solution formed immediately, followed by the precipitation of a yellow-green solid. The mixture was cooled to −35 °C, resulting in further precipitate. The solvent was decanted, and the precipitate dissolved in minimal diethyl ether and crystallized by slow pentane diffusion into the supernatant. (25 mg, 63%) ^{1}H-NMR (C$_6$D$_6$): δ 82.2 (s, 3H, PhB(^{Mes}*Im*)$_3$FeNPPh$_3$); 68.8 (s, 3H, PhB(^{Mes}*Im*)$_3$FeNPPh$_3$); 53.7 (s, 2H, *Ph*B(^{Mes}Im)$_3$FeNPPh$_3$); 49.0 (s, 6H, PhB(^{Mes}Im)$_3$FeN*PPh$_3$*); 25.9 (s, 2H, *Ph*B(^{Mes}Im)$_3$FeN-PPh$_3$); 23.0 (s, 6H, PhB(^{Mes}Im)$_3$FeN*PPh$_3$*); 22.1 (s, 1H, *Ph*B(^{Mes}Im)$_3$FeNPPh$_3$); 8.4 (s, 3H, PhB(^{Mes}Im)$_3$FeN*PPh$_3$*); 3.2 (s, 9H, PhB(^{Mes}Im)$_3$FeNPPh$_3$); −3.0 (s, 6H, PhB(^{Mes}Im)$_3$FeNPPh$_3$); −51.0 (s, 18H, PhB(^{Mes}Im)$_3$FeNPPh$_3$). μ_{eff} = 5.2(3) μ_B.

128   6   PhB(^{Mes}Im)$_3^-$: Spin Crossover in a Four-Coordinate Iron(II) Complex

Elemental Analysis (%) for $C_{60}H_{59}BFeN_7P\cdot0.5\ C_4H_{10}O$ calcd. C 73.52, H 6.37, N 9.68; obsd. C 73.22, H 6.15, N 9.90.

6.4.3 Magnetic Susceptibility Measurements

As shown in Fig. 6.1, at 300 K, $\chi_M T = 3.63\ cm^3\ mol\ K^{-1}$, consistent with a high spin $S = 2$ state with $g = 2.2$. Here, the presence of a g value greater than 2.0 likely arises due to the presence of unquenched orbital angular momentum associated with the $e_{(a)}^3$ orbital set. As the temperature is lowered, $\chi_M T$ remains relatively constant to below 150 K, before undergoing a gradual decline then precipitous drop at 81 K. Below 50 K, the data decrease only very slightly, with an average value in the temperature range 10–50 K of $\chi_M T = 0.42\ cm^3\ mol\ K^{-1}$. Below 10 K, the data undergo a rapid decline, likely a result of Zeeman and zero-field splitting. Measuring the susceptibility as the temperature is increased from 2 to 300 K leads to a $\chi_M T$ curve essentially identical to that obtained from decreasing the temperature from 300 to 2 K. A very small discrepancy in the two datasets below 30 K likely results from slight torquing of the crystallites within the 1 T applied field. Notably, no hysteresis is evident at the applied sweep-rates near the transition, dictated by the temperature intervals over which the data were collected.

The crossover phenomenon associated with **1** was modeled using a simple Boltzmann distribution of spin states employing the following two equations [13]:

$$x = 1/[1 + \exp(\Delta H/R)(1/T - 1/T_C)] \tag{6.1}$$

$$x = [\chi T - (\chi T)_{LS}]/[(\chi T)_{HS} - (\chi T)_{LS}] \tag{6.2}$$

where x is the molar fraction of high-spin molecules, ΔH is the change in enthalpy associated with the spin state transition, R is the molar gas constant ($0.695\ cm^{-1}$ K^{-1}), and T_C is the critical temperature. The corresponding least squares fit to the warming-mode data in the temperature range 20–300 K, with $(\chi T)_{LS}$ and $(\chi T)_{HS}$ held constant at 0.42 and 3.63 $cm^3\ mol\ K^{-1}$, respectively, provides values of $\Delta H = 1190\ cm^{-1}$ and $T_C = 81$ K. While this simple Boltzmann model reproduces the critical temperature and slope of the data very near the transition, it fails to provide a close fit to the data at temperatures around the cusps of the transition (Scheme 6.1). This discrepancy may be the result of cooperative interactions between neighboring molecules. Nevertheless, attempts to fit the data using models that take such interactions into account, in particular the regular solution and domain models, do not provide fits of significantly higher quality.

To address the drop in $\chi_M T$ below ca. 20 K, the data were modeled using MAGPACK [14], taking into consideration Zeeman splitting and axial zero-field splitting, according to the following Hamiltonian:

$$\widehat{H} = D\widehat{S}_Z^2 + g_{iso}\mu_B \mathbf{S} \cdot \mathbf{H}$$

6.4 Experimental

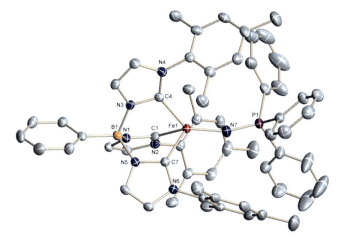

Fig. 6.7 Molecular structure of **1** at 150 K. Thermal ellipsoids shown at 50%, hydrogen atoms omitted for clarity

Table 6.2 Electronic energy differences between the high spin and low spin states as determined by DFT methods

Basis set	ΔE (HS−LS)
B3LYP	−7.6
BPW91	18.0

Energies are at 0 K in kcal mol^{-1}

Simulating the data for an 11.6% population of an $S = 2$ state with $g = 2.2$ gives an axial zero-field splitting parameter of $|D| = 5(2)$ cm^{-1}. Here, the error accounts for the slight difference in the warming- and cooling-mode datasets.

6.4.4 Computational Details

Density functional calculations were performed at the Molecular Graphics Laboratory, University of California-Berkeley. Full geometry optimizations were performed on the complete molecule [PhB(^{Mes}Im)$_3$Fe–N=PPh$_3$] using the Gaussian 09 software package [17] using both the hybrid B3LYP and exchange–correlation BPW91 functionals. The 6-311G(df) basis set was used for the iron atom and 6-31G(d) was used for all other atoms. Optimizations were performed for the $S = 0$ and $S = 2$ spin states and frequency calculations were used to confirm minima. The optimized energies (Table 6.2) show that the BPW91 basis set predicts the correct ground state, whereas the B3LYP basis set does not. Selected bond distances and angles are compared with the crystallographic data in Table 6.3.

Table 6.3 Selected bond distances [Å] and bond angles [°] for the two spin states of **1** determined by X-ray crystallography and DFT calculations

	High spin ($S = 2$)				Low spin ($S = 0$)		
	X-ray		DFT		X-ray	DFT	
	150 K	100 K	BPW91	B3LYP	30 K	BPW91	B3LYP
Fe1–N7	1.855(2)	1.853(2)	1.843	1.868	1.807(2)	1.789	1.825
Fe1–C4	2.079(2)	2.071(2)	2.040	2.132	1.939(2)	1.882	1.928
Fe1–C1	2.086(2)	2.073(2)	2.037	2.117	1.954(2)	1.87	1.929
Fe1–C7	2.089(2)	2.076(2)	2.087	2.106	1.951(2)	1.882	1.910
P1–N7	1.524(2)	1.525(2)	1.565	1.542	1.549(2)	1.605	1.571
P1–N7–Fe1	162.0(2)	161.4(2)	174.11	174.14	171.0(2)	178.78	178.59
C4–Fe1–C1	87.36(9)	87.31(9)	90.99	89.08	89.20(9)	88.27	86.80
C7–Fe1–C1	91.11(9)	96.27(9)	93.40	90.06	85.28(9)	88.31	88.14
C4–Fe1–C7	86.24(9)	91.04(9)	90.59	88.23	86.93(9)	86.86	88.11

6.4.5 X-ray Crystal Structure Determination Details

X-ray crystal structure determinations were performed on two crystals derived from two independently synthesized and crystallized samples of the compound at 30 and 150 K in each case. Here, we report only the results of the best data set obtained at a given temperature.

Brownish orange blocks of [PhB(^{Mes}Im)$_3$Fe–N=PPh$_3$] · toluene at 30 K (**1** toluene at 30 K) were grown from a toluene solution of the complex at −37 °C. Orange blocks of [PhB(^{Mes}Im)$_3$Fe–N=PPh$_3$] · toluene at 150 K (**1** · toluene at 150 K) were grown from a toluene solution of the complex at −37 °C. Suitable single crystals of the compound were embedded in protective perfluoropolyalkylether oil and quickly transferred to the cold gas stream of the diffractometer. Intensity data were collected at 30 and at 150 K on a Bruker Smart APEX2 diffractometer (MoK$_\sigma$ radiation, $\lambda = 0.71073$ Å, graphite monochromator). Lorentz and polarization effects were taken into account during data reduction, semiempirical absorption corrections were performed on the basis of multiple scans using *SADABS* [22]. All structures were solved by direct methods and refined by full-matrix least-squares procedures on F^2 using *SHELXTL NT* 6.12 [23] (Table 6.4).

[PhB(^{Mes}Im)$_3$Fe–N=PPh$_3$] · toluene at 30 K (**1** · toluene at 30 K) crystallizes with one molecule of toluene that is disordered. Two alternative orientations were refined that are occupied by 66.3(5) and 33.7(5)% for atoms C101–C107 and C111–C117, respectively. SIMU, ISOR, and SAME restraints were applied in the refinement of the disordered toluene molecule. [PhB(^{Mes}Im)$_3$Fe–N=PPh$_3$] · toluene at 150 K (**1** · toluene at 150 K) crystallizes with one molecule of toluene that is disordered. Two alternative orientations were refined that are occupied by 62.6(9) and 37.4(9)% for atoms C101–C107 and C111–C117, respectively. DFIX, SIMU, ISOR, and SAME restraints were applied in the refinement of the disordered toluene molecule (Fig. 6.7).

6.4 Experimental

Table 6.4 Crystallographic data, data collection, and refinement details of [PhB(^{Mes}Im)$_3$Fe–N = PPh$_3$] · toluene at 30 K (**1** · toluene at 30 K) and [PhB(^{Mes}Im)$_3$Fe–N = PPh$_3$] · toluene at 150 K (**1** · toluene at 150 K)

	1 · toluene at 30 K	**1** · toluene at 150 K
Empirical formula	C$_{67}$H$_{67}$BFeN$_7$P	C$_{67}$H$_{67}$BFeN$_7$P
Mol. weight	1067.91	1067.91
Crystal size [mm^3]	0.46 × 0.19 × 0.15	0.28 × 0.21 × 0.10
Temperature [K]	30	150
Crystal system	Monoclinic	Monoclinic
Space group	$P2_1/n$ (no. 14)	$P2_1/n$ (no. 14)
a [Å]	21.820(2)	22.113(2)
b [Å]	12.052(2)	12.1722(7)
c [Å]	22.574(2)	22.519(2)
α [°]	90	90
β [°]	113.39(1)	112.62(1)
γ [°]	90	90
V [Å^3]	5448.4(8)	5595.3(6)
Z	4	4
ρ_{calc} [g cm^{-3}]	1.302	1.268
μ [mm^{-1}]	0.357	0.347
F (000)	2256	2256
T_{min}; T_{max}	0.630; 0.746	0.664; 0.746
2θ interval [°]	$5.1 \leq 2\theta \leq 54.2$	$5.1 \leq 2\theta \leq 54.2$
Collected reflections	52492	53818
Independent reflections; R_{int}	11734; 0.0547	11803; 0.0403
Observed reflections [$I \geq 2\sigma(I)$]	9324	8934
No. refined parameters	728	744
wR_2 (all data)	0.1302	0.1376
R_1 [$I \geq 2\sigma(I)$]	0.0471	0.0469
GooF on F^2	1.045	1.066
$\Delta\rho_{max/min}$	0.772/−0.717	0.693/−0.542

All non-hydrogen atoms were refined with anisotropic displacement parameters. All hydrogen atoms were placed in positions of optimized geometry, their isotropic displacement parameters were tied to those of the corresponding carrier atoms by a factor of either 1.2 or 1.5.

Acknowledgments Text, schemes, and figures of this chapter, in part, are reprints of the materials published in Scepaniak et al. [16]. The dissertation author was the tertiary researcher and author. The co-authors listed in the publication also participated in the research. The permission to reproduce the paper was granted by the American Chemical Society. Copyright 2011, American Chemical Society.

References

1. P. Gütlich, H.A. Goodwin, Top. Curr. Chem. **233**, 1 (2004)
2. P. Gütlich, Y. Garcia, H.A. Goodwin, Chem. Soc. Rev. **29**, 419 (2000)
3. P. Gütlich, A. Hauser, H. Spiering, Angew. Chem. Int. Ed. **33**, 2024 (1994)
4. J.A. Real, A.B. Gaspar, M.C. Munoz, Dalton Trans. 2062 (2005)
5. J.-F. Létard, P. Guionneau, L. Goux-Capes, Top. Curr. Chem. **235**, 221 (2004)
6. M.A. Halcrow, Polyhedron **26**, 3523 (2007)
7. J. Li, R.L. Lord, B.C. Noll, M.-H. Baik, C.B. Schulz, W.R. Scheidt, Angew. Chem. Int. Ed. **47**, 10144 (2008)
8. D.M. Jenkins, J.C. Peters, J. Am. Chem. Soc. **127**, 7148 (2005)
9. D.M. Jenkins, J.C. Peters, J. Am. Chem. Soc. **125**, 11162 (2003)
10. J.J. Scepaniak, T.D.Harris, C.S.Vogel, J.Sutter, K.Meyer, J.M. Smith, J. Am. Chem. Soc. **133**, 3824 (2011)
11. J.J. Scepaniak, J.A. Young, R.P. Bontchev, J.M. Smith, Angew. Chem. Int. Ed. **48**, 3158 (2009)
12. J.J. Scepaniak, M.D. Fulton, R.P. Bontchev, E.N. Duesler, M.L. Kirk, J.M. Smith, J. Am. Chem. Soc. **130**, 10515 (2008)
13. O. Kahn, *Molecular Magnetism* (Wiley-VCH, New York, 1993)
14. J.J. Borrás-Almenar, J.M. Clemente-Juan, E. Coronado, B.S. Tsukerblat, J. Comput. Chem. **22**, 985 (2001)
15. I. Nieto, F. Cervantes-Lee, J.M. Smith, Chem. Commun. (Cambridge, UK) 3811 (2005)
16. J.J. Scepaniak, C.S. Vogel, M.M. Khusniyarov, F.W. Heinemann, K. Meyer, J.M. Smith, Science **331**, 1049 (2011)
17. Gaussian 09, Revision A1, M.J. Frisch, G.W. Trucks, H. B Schlegel, G. E. Scuseria, M.A. Robb, J.R. Cheeseman, G Scalmani, V. Barone, B. Mennucci, G. A Petersson, H. Nakatsuji, M. Caricato, X. Li, H.P. Hratchian, A.F Izmaylov, J Bloino, G. Zheng, J.L. Sonnenberg, M. Hada, M. Ehara, K. Toyota, R. Fukuda, J. Hasegawa, M. Ishida, T. Nakajima, Y. Honda, O. Kitao, H. Nakai, T. Vreven, J A. Montgomery Jr., J. E. Peralta, F. Ogliaro, M. Bearpark, J.J. Heyd, E. Brothers, K.N. Kudin, V.N. Staroverov, R. Kobayashi, J. Normand, K. Raghavachari, A. Rendell, J.C. Burant, S. S. Iyengar, J. Tomasi, M. Cossi, N. Rega, N.J. Millam, M. Klene, J.E. Knox, J. B. Cross, V. Bakken, C. Adamo, Jaramillo, J.; Gomperts, R.; Stratmann, R. E.; Yazyev, O.; Austin, A. J.; Cammi, R.; Pomelli, C.; Ochterski, J. W.; Martin, R. L.; Morokuma, K.; Zakrzewski, V. G.; Voth, G. A.; Salvador, P.; Dannenberg, J. J.; Dapprich, S.; Daniels, A. D.; Farkas, Ö.; Foresman, J. B.; Ortiz, J. V.; Cioslowski, J.; Fox, D. J. Gaussian, Inc., Wallingford CT, (2009)
18. S. Alvarez, J. Circa, Angew. Chem. Int. Ed. **45**, 3012 (2006)
19. E. Tangen, J. Conradie, A. Ghosh, J. Comput. Theory Comput. **3**, 448 (2007)
20. S.D. Brown, J.C. Peters, J. Am. Chem. Soc. **127**, 1913 (2005)
21. M.V. Baker, L.D. Field, T.W. Hambley, Inorg. Chem. **27**, 2872 (1988)
22. SADABS 2008/1; Bruker AXS, Inc., (2008)
23. SHELXTL NT 6.12; Bruker AXS, Inc., (2002)